Influenza Virus and Vaccination

Influenza Virus and Vaccination

Special Issue Editors

Marta L. DeDiego
Aitor Nogales

MDPI • Basel • Beijing • Wuhan • Barcelona • Belgrade

Special Issue Editors
Marta L. DeDiego
Department of Molecular and Cell Biology, Centro Nacional de Biotecnología (CNB-CSIC)
Spain

Aitor Nogales
The Center for Animal Health Research (CISA-INIA)
Spain

Editorial Office
MDPI
St. Alban-Anlage 66
4052 Basel, Switzerland

This is a reprint of articles from the Special Issue published online in the open access journal *Pathogens* (ISSN 2076-0817) from 2019 to 2020 (available at: https://www.mdpi.com/journal/pathogens/special_issues/InfluenzaVirus).

For citation purposes, cite each article independently as indicated on the article page online and as indicated below:

LastName, A.A.; LastName, B.B.; LastName, C.C. Article Title. *Journal Name* **Year**, *Article Number*, Page Range.

ISBN 978-3-03928-817-5 (Pbk)
ISBN 978-3-03928-818-2 (PDF)

Contents

About the Special Issue Editors

Marta L. DeDiego I finished my BS degree in Biochemistry at the Universidad Autónoma of Madrid (Spain), in 2003, and received my PhD from the same university at the end of 2008, working at the National Center for Biotechnology (CNB-CSIC) in Madrid. Then, I stayed in the same laboratory for a postdoctoral term, and in 2013, I moved to the University of Rochester (USA). Since April, 2018, I am working at the National Center for Biotechnology, in Madrid (Spain), as a Principal Investigator. My research is focused on viral genes and immune responses contributing to the severity of respiratory virus infections and the development of potential vaccines against these pathogens. We have generated the second and first infectious cDNA clones for the deadly Severe Acute Respiratory Syndrome Coronavirus (SARS-CoV) and Middle East Respiratory Syndrome Coronavirus (MERS-CoV), respectively. We established the molecular basis of coronavirus or influenza virulence and generated attenuated viruses as vaccine candidates, showing that coronavirus genes such as nsp1 and the envelope (E) gene or influenza genes such as NS1 or PA-X modulate innate immune responses after infection and contribute to viral pathogenesis, and demonstrating that viruses lacking these genes (coronavirus) or with altered functional properties (influenza) are attenuated and promising vaccine candidates. In fact, a SARS-CoV lacking the E gene was fully attenuated and fully protected in animal models, creating the first recombinant, live-attenuated SARS vaccine candidate. In addition, we found mutations in the influenza HA and NA proteins contributing to antibody evasion recognition. Furthermore, we showed that passaging the influenza vaccine strain virus under immune pressure provided by human sera selects for viruses incorporating mutations in the HA protein matching the mutations present in circulating viruses, making this strategy informative to update the annual influenza vaccine reformulation. Using the knowledge about viral pathogenesis and host inflammatory response pathways garnered in my research, the current strategy of my team is to identify viral and host factors contributing to the severity of influenza and coronavirus respiratory infections and use this knowledge for the development of new approaches to prevent (vaccines) or treat (antivirals) these viral infections.

Aitor Nogales I finished my BS degree in Biochemistry at the Universidad Autonoma de Madrid (Spain) and received my PhD (2012) from the same university, working in the laboratory of Coronavirus at the National Center for Biotechnology (CNB-CSIC) (Spain). The topic of my PhD was "Coronavirus replication mechanism. Identification of host and viral proteins involved in coronavirus replication" and I obtained the highest grade of "Sobresaliente cum laude" (outstanding honor). In 2013, I moved to the University of Rochester (USA) as a postdoctoral fellow, and later, due to my professional merits I was promoted to Research Assistant Professor. Since 2019, the prestigious Ramón y Cajal (R&C) award has allowed my incorporation as Principal Investigator to the Center for Animal Health Research (Centro de Investigación en Sanidad Animal; CISA), part of the National Institute for Agricultural and Food Research and Technology (Instituto Nacional De Investigaciones Agrarias; INIA), which is a multidisciplinary research center with infrastructures to address research in human and animal health, under high Biosafety facilities. My research interest has been focused on the molecular biology, virus-host interaction, and pathogenesis of multiple RNA and DNA viruses as well as the development of novel therapeutic strategies to prevent (vaccines) or combat (antiviral therapies) viral infections. I have developed an extensive research activity mainly, in understanding the molecular mechanisms responsible for influenza virus infection and the development of novel approaches against human, canine, equine, and avian influenza viruses.

pathogens

MDPI

Editorial

Influenza Virus and Vaccination

Aitor Nogales [1,*] and Marta L. DeDiego [2,*]

1 Center for Animal Health Research, INIA-CISA, 28130 Madrid, Spain
2 Department of Molecular and Cell Biology, Centro Nacional de Biotecnología (CNB-CSIC), Campus Universidad Autónoma de Madrid, 28049 Madrid, Spain
* Correspondence: nogales.aitor@inia.es (A.N.); Marta.Lopez@cnb.csic.es (M.L.D.)

Received: 11 March 2020; Accepted: 11 March 2020; Published: 17 March 2020

Abstract: Influenza virus infections represent a serious public health problem causing contagious respiratory disease and substantial morbidity and mortality in humans, resulting in a considerable economic burden worldwide. Notably, the number of deaths due to influenza exceeds that of any other known pathogen. Moreover, influenza infections can differ in their intensity, from mild respiratory disease to pneumonia, which can lead to death. Articles in this Special Issue have addressed different aspects of influenza in human health, and the advances in influenza research leading to the development of better therapeutics and vaccination strategies, with a special focus on the study of factors associated with innate or adaptive immune responses to influenza vaccination and/or infection.

Keywords: Influenza virus; influenza vaccine; vaccination; pandemic; immune response; innate immunity; adaptive immunity; universal vaccines

---

Sangster et al. provide a comprehensive picture of HA-specific antibody response to influenza virus infection, relevant for protecting the host against the infection. Moreover, the authors discuss the importance of the composition of an individual's HA-reactive preexisting memory B cell (MBC), a population that can reflect the imprint of early-life HA exposure, and its role in determining the character of the HA-reactive antibody response. Furthermore, the authors suggest that antibodies resulting from preexisting MBC activation are important regulators of anti-HA antibody production, and play a role in the positive selection of germinal center B cells which are reactive to novel HA epitopes. The understanding of MBC competition, immunodominance hierarchies, and antibody regulation of B cell responses will help to improve influenza vaccine composition and administration approaches [1]. As reviewed by Jiong et al., traditional assays to study complex humoral responses after influenza infection or vaccination, such as HA-specific antibody responses, are limited in scope and too resource-intensive. However, multidimensional assays developed in recent years could overcome these problems by simultaneously measuring antibodies against a large panel of influenza HA proteins in a high throughput assay [2]. Misra and Nayak have highlighted the importance of vaccinating children and pregnant women against influenza, since influenza virus infection is responsible for significant morbidity and mortality in these populations. Disturbingly, the authors indicated that despite the benefits of the influenza vaccine, vaccination rates around the world remain well below targets. The constantly changing HA antigenicity of the influenza virus, along with the complexity of serological responses induced by influenza infections in the immune system, muddies efforts to interpret serology testing results or develop more effective vaccines [3]. In order to improve the safety of the live-attenuated influenza vaccine (LAIV) and make it available to a broader population (it is currently not recommended for children under the age of two, immunocompromised individuals, the elderly, and pregnant adults), Hilimire et al. have demonstrated that the influenza A virus master donor virus (MDV) A/Ann Arbor/6/60 H2N2 LAIV can inhibit host gene expression using both the PA-X and NS1 viral proteins. Furthermore, they show that by removing PA-X, the replication of the

MDV LAIV is decreased in a mouse model, while maintaining full protective efficacy, demonstrating a broadly applicable strategy of tuning the amount of host antiviral responses induced by the IAV MDV for the development of improved and safer LAIVs [4].

Topham et al. discuss recent advances in an important branch of adaptive immune responses, the tissue resident memory (TRm) CD_8 T cells, which comprise a cell population that forms in peripheral, nonlymphoid tissue after infection, and that do not recirculate into the bloodstream or other tissues. This cell population has been shown to be important against secondary encounters with a previously seen pathogen. However, many questions remain regarding our understanding of this unique cell subset and its role during influenza infections [5]. Previously, Sant's laboratory has demonstrated that CD_4 T cells specific for epitopes derived from HA are the most effective in providing help for the HA-specific B cell responses to infection and vaccination. In this special issue, Zackery et al. ask whether HA epitopes recognized by CD_4 T cells in the primary response to infection are equally distributed across the HA protein. Using mice, their studies revealed that the HA-specific CD_4 T cell epitopes cluster in two distinct regions of HA, which could be important in the development of universal vaccines against influenza [6].

Nogales and DeDiego discuss the importance of human genome polymorphisms for the susceptibility of some individuals to suffer more severe symptoms after influenza infections and for vaccine effectiveness. Notably, the knowledge and analysis of host genome variability will be a valuable tool with which to predict the outcomes of viral diseases and of prophylactic or therapeutic interventions, including vaccines and drugs [7].

Author Contributions: A.N. and M.L.D. conceptualized and wrote the manuscript. All authors have read and agreed to the published version of the manuscript.

Funding: This work was supported with funds from Comunidad de Madrid, Spain, reference 2017-T1/BMD-5155 and the Spanish Ministry of Science, Innovation and Universities (RTI-2018-094213-A-I00) to M.L.D and the "Ramon y Cajal" Fellowship RYC-2017 from Spanish Ministry of Economy, Industry and Competitiveness to A.N.

Acknowledgments: We thank all authors who have contributed to this special issue.

Conflicts of Interest: The authors declare no conflict of interest.

References

1. Sangster, M.Y.; Nguyen, P.Q.T.; Topham, D.J. Role of Memory B Cells in Hemagglutinin-Specific Antibody Production Following Human Influenza A Virus Infection. *Pathogens* **2019**, *8*, 167. [CrossRef] [PubMed]
2. Wang, J.; Wiltse, A.; Zand, M.S. A Complex Dance: Measuring the Multidimensional Worlds of Influenza Virus Evolution and Anti-Influenza Immune Responses. *Pathogens* **2019**, *8*, 238. [CrossRef] [PubMed]
3. Misra, R.S.; Nayak, J.L. The Importance of Vaccinating Children and Pregnant Women against Influenza Virus Infection. *Pathogens* **2019**, *8*, 265. [CrossRef] [PubMed]
4. Hilimire, T.A.; Nogales, A.; Chiem, K.; Ortego, J.; Martinez-Sobrido, L. Increasing the Safety Profile of the Master Donor Live Attenuated Influenza Vaccine. *Pathogens* **2020**, *9*, 86. [CrossRef] [PubMed]
5. Topham, D.J.; Reilly, E.C.; Emo, K.L.; Sportiello, M. Formation and Maintenance of Tissue Resident Memory CD8+ T Cells after Viral Infection. *Pathogens* **2019**, *8*, 196. [CrossRef] [PubMed]
6. Knowlden, Z.A.G.; Richards, K.A.; Moritzky, S.A.; Sant, A.J. Peptide Epitope Hot Spots of CD4 T Cell Recognition Within Influenza Hemagglutinin During the Primary Response to Infection. *Pathogens* **2019**, *8*, 220. [CrossRef] [PubMed]
7. Nogales, A.; M, L.D. Host Single Nucleotide Polymorphisms Modulating Influenza A Virus Disease in Humans. *Pathogens* **2019**, *8*, 168. [CrossRef]

pathogens

MDPI

Article

Peptide Epitope Hot Spots of CD4 T Cell Recognition Within Influenza Hemagglutinin During the Primary Response to Infection

Zackery A. G. Knowlden, Katherine A. Richards, Savannah A. Moritzky and Andrea J. Sant *

David H. Smith Center for Vaccine Biology and Immunology, Department of Microbiology and Immunology, University of Rochester Medical Center, Rochester, NY 14642, USA; Zackery_Knowlden@URMC.Rochester.edu (Z.A.G.K.); Katherine_skelly@urmc.rochester.edu (K.A.R.); Savannah_Moritzky@URMC.Rochester.edu (S.A.M.)

* Correspondence: Andrea_sant@urmc.rochester.edu

Received: 2 October 2019; Accepted: 29 October 2019; Published: 5 November 2019

Abstract: Antibodies specific for the hemagglutinin (HA) protein of influenza virus are critical for protective immunity to infection. Our studies show that CD4 T cells specific for epitopes derived from HA are the most effective in providing help for the HA-specific B cell responses to infection and vaccination. In this study, we asked whether HA epitopes recognized by CD4 T cells in the primary response to infection are equally distributed across the HA protein or if certain segments are enriched in CD4 T cell epitopes. Mice that collectively expressed eight alternative MHC (Major Histocompatibility Complex) class II molecules, that would each have different peptide binding specificities, were infected with an H1N1 influenza virus. CD4 T cell peptide epitope specificities were identified by cytokine EliSpots. These studies revealed that the HA-specific CD4 T cell epitopes cluster in two distinct regions of HA and that some segments of HA are completely devoid of CD4 T cell epitopes. When located on the HA structure, it appears that the regions that most poorly recruit CD4 T cells are sequestered within the interior of the HA trimer, perhaps inaccessible to the proteolytic machinery inside the endosomal compartments of antigen presenting cells.

Keywords: influenza; CD4 T cell; epitopes; hemagglutinin

1. Introduction

Influenza hemagglutinin (HA) is a major target of influenza vaccination because of the importance that this molecule plays in mediating binding and infection of influenza virus on host cells. Because of this function, antibodies elicited by most influenza vaccine strategies have the potential to provide sterilizing immunity to influenza infection in the host (reviewed in [1–7]). CD4 T cells specific for HA are also critical in the protective antibody responses to infection and vaccination because of their role in the germinal center response, high affinity antibody production and B cell memory (reviewed in [8–10]). Our studies show that in human vaccine responses to influenza, the serum antibody responses to HA are correlated with the elicitation of CD4 T cells specific for HA peptide epitopes and not those from other viral proteins contained in the vaccine, such as NP (nucleoprotein) [11]. Similarly, in animal models of influenza infection, recall of memory CD4 T cells specific for HA, but not NP, is associated with neutralizing antibody response to HA [12]. The preceding studies suggest that, although the human CD4 T cell repertoire is highly diverse with respect to reactivity to influenza viral proteins [13–21], understanding the specificity of CD4 T cells towards the HA protein is particularly important.

There have been a number of studies that have reported individual HA epitopes that are recognized by CD4 T cells in both human and mouse models of infection and vaccination (reviewed in [22,23]). However, published studies have often been limited to the analyses of particular regions of HA, chosen based on common sequences in diverse subtypes of influenza [24], or shared epitopes between seasonal and potentially pandemic strains of influenza [16,20,25–29]. A subset of peptides in HA have also been studied because they are presented by particular allelic forms of human HLA (Human Leukocyte Antigen) class II molecules that are well-suited for class II-peptide tetramer analyses [14,28,30–32]. Because of our interest in the full potential repertoire of influenza-specific CD4 T cells, our laboratory has used unbiased epitope scanning methods to identify the entire array of peptides recognized by CD4 T cells in the primary response in mice to influenza infection or vaccination. These studies have involved a diverse set of mouse strains, chosen in part to explore the role of MHC (Major Histocompatibility Complex) class II polymorphism in selection of the immunodominance hierarchy [33,34], the relationship between epitope selection in CD4 T cell responses to infection vs. vaccination [35], CD4 T cell epitope selection during homing to the lung after infection [36,37], partitioning of CD4 T cells into follicular helper cells versus effector cell lineages after infection and vaccination [38], and epitope mapping for identification of human CD4 T cell epitopes via HLA-DR transgenic mice [39,40]. Collectively, these studies have revealed a tremendous breadth in the CD4 T cell response to influenza that includes specificities derived from HA, NA, M1, NP, and the polymerase proteins.

In the study reported here, we analyzed the peptide epitope distribution of HA-specific CD4 T cells elicited by influenza infection from a panel of independent strains of mice. These mice were chosen because, based on unique MHC class II molecules that they express, they will each select an independent CD4 T cell peptide repertoire from the HA protein. Our studies revealed that the epitopes selected for CD4 T cells in the primary response are localized to a limited number of "hot spots" within the HA protein and that there are corresponding "dead zones" that recruit very few CD4 T cells. The basis of the immunodominance in the influenza HA protein is not fully understood at this time but is important to understand in developing vaccine strategies for influenza virus.

2. Results

2.1. HA Epitope Dominance in the Primary Response to Influenza Infection

To map individual epitopes from the influenza HA protein, common inbred mice expressing different murine MHC class II molecules (thus having the potential to present different HA-derived peptides) or transgenic mice, expressing human HLA-DR or HLA-DQ molecules, were infected with A/New Caledonia/20/99 H1N1 virus, a recent human isolate which replicates in mice [41] but that is not highly pathogenic in any of the strains tested. At day 10–13 post-infection, peptide epitopes recognized by the elicited CD4 T cells were identified by a sequential method involving a peptide pooling matrix, which we described in detail previously [39,40]. In this method, single overlapping peptides representing the entire translated HA protein are pooled in sets of 9–10 different peptides and arrayed in a matrix strategy originally described by Tobery and colleagues for identification of T cell specificities [42]. Spleen cells from infected mice were used as a source of primed CD4 T cells, enriched by negative paramagnetic bead isolation, as described in Materials and Methods. CD4 T cells were tested for reactivity to the HA peptide pools that were arrayed in a matrix design and quantified using cytokine EliSpot assays. IL-2 producing CD4 T cells were used to quantify reactivity to influenza epitopes from spleen because in some strains of mice, this is the dominant cytokine expressed particularly in the spleen ([41] and data not shown). In addition, in contrast to cells localized to the lung, that are enriched for IFN-γ [36,37], IL-2 is reliably expressed by CD4 T cells elicited by infection that localize to secondary lymphoid tissue [34]. Use of peptides and splenic CD4 T cells allowed all potential HA epitopes recognized by the host CD4 T cells to be presented by APC and available to recall the primed CD4 T cells. The primed CD4 T cells and syngeneic splenocytes from naïve mice or MHC class II expressing fibroblasts (for HLA-DR1 transgenic mice), used as a source of APC, were co-incubated with the peptide pools for 16–20 h. Pools tested as negative were eliminated from further analyses. Candidate peptides in the positive rows and columns were then tested as single peptides. Through this sequential, iterative process, the single CD4 T cell peptide epitopes derived from HA were identified.

Supplemental Table S1 lists each of the peptides in the overlapping peptide library that was used in the initial epitope screening for each mouse strain. Supplemental Table S2 shows the peptide specificities of all the mouse strains that were analyzed with the indicated average cytokine spot count from at least three replicate experiments. We classified peptides that elicited at least 125 cytokine-producing cells as positive for this study because these were the most reliably quantified in replicate, independent assays. These curated data are shown in Table 1, which also denotes the mouse strain and MHC class II haplotype and allelic forms of class II expressed. Because of the large range in responses to different peptide epitopes, we subdivided the positive epitopes into two groups: subdominant (125–300 cytokine-producing cells) and dominant (>300 cytokine-producing cells). The typical range in HA-reactive epitope-specific CD4 T cells in any given mouse strain ranged from 125–400 spots, although an occasional epitope recruited more than 600 cytokine-producing CD4 T cells.

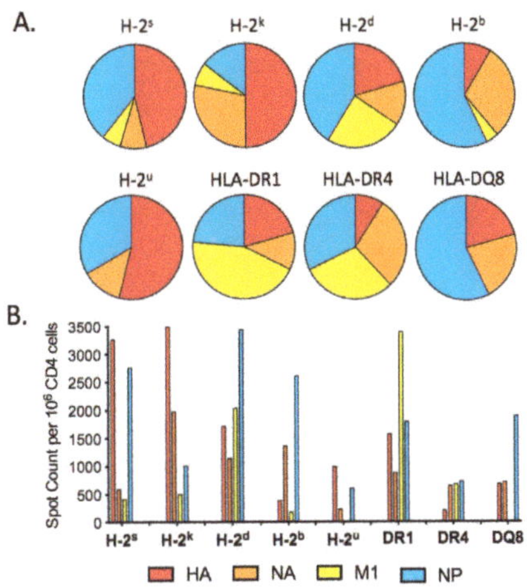

Figure 1. Pattern of CD4 T cell reactivity to influenza viral proteins following primary infection. Influenza reactivity to influenza viral proteins following primary infection with A/New Caledonia/20/99 was measured by IL-2 EliSpot assays as spots per million CD4 T cells to HA (red), NA (orange), M1 (yellow), and NP (blue), where the total number of CD4 T cells producing IL-2 specific for defined epitopes within each viral protein were summed. In panel **A** the immunodominance hierarchies for each mouse strain are illustrated as pie charts, with the fraction of the response dedicated to each viral protein is indicated. Responses for the M1 protein were not assessed in the H-2u and HLA-DQ8 mouse models and therefore their immunodominace hierarchies as illustrated are limited to HA, NA and NP. In panel **B** the total spot counts are indicated as bar graphs for each MHC haplotype in the strains of mice analyzed.

Table 1. Mouse strain dependent immunodominance hierarchy in CD4 T cell epitopes specific for HA from A/New Caledonia/20/99. Each mouse strain with HA epitopes >125 spots/million CD4 T cells is shown. Also indicated for each strain is the MHC haplotype and MHC class II molecules expressed. On the far right is the average cytokine spots/million for each epitope from at least 3 independent experiments. These epitopes were those selected for the data shown in Figures 1 and 2. The peptides in bold represent those with 300 spots/million.

Mouse Strain	MHC Haplotype	Peptide	Amino Acids	Sequence	Spots Per 10^6 Cells
A/J	H-2^a	**HA p21**	**120–136**	**120 EQLSSVSSFERFEIFPK 136**	**474**
		HA p30	**174–190**	**174 YPNLSKSYVNNKEKEVL 190**	**374**
		HA p37	215–231	215 VSVVSSHYSRRFTPEIA 231	236
		HA p56	328–344	328 LRMVTGLRNIPSIQSRG 344	271
		HA p61	**358–374**	**358 TGMVDGWYGYHHQNEQG 374**	**507**
		HA p64	**375–391**	**375 SGYAADQKSTQNAINGI 391**	**351**
		HA p68	**398–414**	**398 VIEKMNTQFTAVGKEFN 414**	**640**
B10.PL	I-A^u	HA p25	144–160	144 TVTGVSASCSHNGKSSF 160	124
		HA p54	**316–332**	**316 IGECPKYVRSAKLRMVT 332**	**369**
		HA p55	322–338	322 YVRSAKLRMVTGLRNIP 338	289
HLA-DQ8	DQ8	HA p14	78–94	78 ILGNPECELLISKESWS 94	136
		HA p48	280–296	280 GIITSNAPMDECDAKCQ 296	155
HLA-DR1	DR1	HA p27	156–172	156 GKSSFYRNLLWLTGKNG 172	167
		HA p28	162–178	162 RNLLWLTGKNGLYPNLS 178	163
		HA p35	203–219	203 NQRALYHTENAYVSVVS 219	159
		HA p74	434–450	434 IWTYNAELLVLLENERT 450	128
		HA p75	440–456	440 ELLVLLENERTLDFHDS 456	146
BALB/c	H-2^d	HA p21	120–136	120 EQLSSVSSFERFEIFPK 136	146
		HA p22	126–142	126 SSFERFEIFPKESSWPN 142	151
		HA p37	215–231	215 VSVVSSHYSRRFTPEIA 231	130
		HA p66	386–402	386 NAINGITNKVNSVIEKM 402	124
SJL	H-2^s	**HA p21**	**120–136**	**120 EQLSSVSSFERFEIFPK 136**	**306**
		HA p22	**126–142**	**126 SSFERFEIFPKESSWPN 142**	**663**
		HA p23	**132–148**	**132 EIFPKESSWPNHTVTGV 148**	**300**
		HA p25	**144–160**	**144 TVTGVSASCSHNGKSSF 160**	**426**
		HA p28	162–178	162 RNLLWLTGKNGLYPNLS 178	263
		HA p54	316–332	316 IGECPKYVRSAKLRMVT 332	191
		HA p57	334–350	334 LRNIPSIQSRGLFGAIA 350	243
		HA p66	**386–402**	**386 NAINGITNKVNSVIEKM 402**	**428**

It is interesting to note that the number of HA-derived epitopes recognized by CD4 T cells from different strains of mice varied considerably (see Supplemental Table S2). For example, H-2^b mice, expressing I-A^b, recruited very few CD4 T cells specific for HA, all below 125 spots per million, while other mice, including A/J mice, expressing I-A^k and I-E^k class II molecules, elicited CD4 T cells specific for greater than fifteen epitopes in HA, six of which were above the subdominant threshold. Although mice tested in the current studies differ both in background genes and in their MHC class II haplotype, our published studies involving B10.S (H-2^s) and B10 (H-2^b) mice that are identical in non-MHC background genes recruit strikingly different epitope specificities [33,34]. These studies, as well as our unpublished data with many inbred strains of mice, suggest that the major factor determining CD4 T cell epitope specificity and abundance after influenza infection is the MHC class II molecules expressed in the host. All strains of mice used in the current study elicited robust CD4 T cell responses to other viral proteins after influenza infection, including those specific for M1, NA, and NP ([22,34,36–40,43], as shown in Figure 1).

2.2. Distribution of Dominant and Subdominant CD4 T Cell Epitopes in the HA Protein Sequence

After compiling all of the CD4 T cell immunodominance hierarchy data, we examined the distribution of epitopes within the HA sequence that met the threshold of a dominant or subdominant epitopes, shown in Supplemental Table S2. These data are shown in Figure 2A. Interestingly, this data representation shows that the peptide epitopes that elicited CD4 T cells were not distributed equally across the HA protein. Instead, they tended to cluster in several distinct regions, which we have termed "hot spots" of reactivity. The biggest clusters of CD4 T cell epitopes in HA, recognized by multiple strains of mice were clustered between amino acids 120–190 and 316–414. Other segments of HA, including amino acid 1–119, 232–315, and 457–565 were relatively devoid of CD4 T cell epitopes, with just an occasional minor peptide epitope represented in these segments. We have termed these regions "dead zones" of reactivity.

We then analyzed these data in a graphical display (Figure 2B). There are approximately 100 16–17 mer peptides in the overlapping array provided by the supplier, numbered from amino acid 1–94, from amino ('N")- to carboxy (C)-terminus (Supplemental Table S2). The peptides were grouped by each decade of HA amino acid sequence (e.g., peptides 1–10, peptides 11–20) (from N to C terminus), indicated on the X-axis. The total CD4 T cell responses, quantified as cytokine spots per million, reflecting the abundance of HA-reactive CD4 T cells in each strain, are depicted by the height of the "cones" on the Y-axis, with each cone depicted by color based on the strain of mouse and the MHC molecules expressed, shown to the left Z-axis. This display allows ready visualization of the distribution and magnitude of the CD4 T cell epitopes, and their clustering across the HA protein. From depicting the immunodominance data in this way, it is clear that among the eight strains of mice studied, there are dominant regions of CD4 T cell recognition within HA. Peptides contained within decade 21–30 and the two decades 51–70, shown in Figure 2, appear to contain the vast majority of the CD4 T cell epitopes. Conversely, there are extended regions of HA that are relatively devoid of CD4 T cell epitopes, notably the first two decades (1–20) and final two decades. The segments of high and very low reactivity, as well as the segments that have only minor epitopes are indicated on the amino acid sequence of the HA protein in red, dark grey, and cream, respectively, in Figure 3.

2.3. Location of Live and Dead Zones on the HA Structure

From our analysis of the "zonal" positioning of CD4 T cell epitopes, we were able to determine regions of the HA protein that had tendencies that favored or disfavored the generation of peptide epitopes in the context of many different MHC molecules. Accordingly, we sought to understand where the zones of reactivity mapped onto the structure of the hemagglutinin protein (Figure 4A,C,D). For this analysis, we designated regions of reactivity in three categories: "live" (regions of high reactivity, colored in red); "dead" (regions of limited/no reactivity, colored in dark grey); and "minor" (regions of occasional CD4 T cell reactivity, colored in cream). As shown in the HA monomer in Figure 4B, live regions stretched from amino acid 120 to 190, and from 316 to 414. Dead regions run from the amino-terminus to amino acid 119, then from 232 to 315 and from 457 to the carboxy-terminus (position 518). The transmembrane domain and membrane proximal amino acids are not depicted in the structure, but are included in the dead zones. Regions of minor reactivity run from residues 191 to 231, as well as 415 to 456. The site where HA0 is cleaved to form HA1 and HA2 (amino acids 343 and 344) is also indicated.

```
MKAKLLVLLCTFTATYADTICIGYHANNSTDTVDTVLEKNVTVTHSVNLLEDSHNGKLCL  60
LKGIAPLQLGNCSVAGWILGNPECELLISKESWSYIVETPNPENGTCYPGYFADYEELRE  120
QLSSVSSFERFEIFPKESSWPNHTVTGVSASCSHNGKSSFYRNLLWLTGKNGLYPNLSK  179
SYVNNKEKEVLVLWGVHHPPNIGNQRALYHTENAYVSVVSSHYSRRFTPEIAKRPKVRDQ  239
EGRINYYWTLLEPGDTIIFEANGNLIAPWYAFALSRGFGSGIITSNAPMDECDAKCQTPQ  299
GAINSSLPFQNVHPVTIGECPKYVRSAKLRMVTGLRNIPSIOSRGLFGAIAGFIEGGWTG  359
MVDGWYGYHHQNEQGSGYAADQKSTQNAINGITNKVNSVIEKMNTQFTAVGKEFNKLERR  419
MENLNKKVDDGFLDIWTYNAELLVLLENERTLDFHDSNVKNLYEKVKSQLKNNAKEIGNG  479
CFEFYHKCNNECMESVKNGTYDYPKYSEESKLNREKIDGVKLESMGVYQILAIYSTVASS  539
LVLLVSLGAISFWMCSNGSLQCRICI     565
```

Mouse Strain	Epitope (Spots/million CD4 T cells)	
	125-299	≥300
SJL (H-2^s)		
BALB/c (H-2^d)		
A/J (H-2^a)		
HLA-DR1		
B10.PL		

A

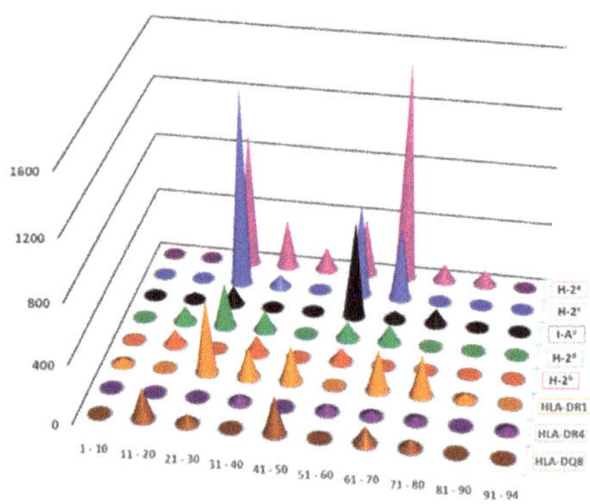

B

Figure 2. CD4 T cell epitopes within the hemmaglutinin protein. Panel (**A**) shows the amino acid sequence from N- to C- terminus including the leader sequence for H1N1 A/New Caledonia/20/99. The underlined segments indicate the amino acids that comprise CD4 T cell peptide epitopes within HA protein that elicit >125 spots/million when measured by cytokine EliSpots. Each color corresponds to a different mouse strain (each with a different MHC haplotype) as indicated, with the darker hue representing ^{3}300 spots/million and the lighter hue representing 125–299 spots/million. Shown to the lower right is the key to the strain designation and color indicators. Responses from HLA-DR4 and C57BL/10 mice are not included in this illustration because their epitopes were below the positive threshold. In Panel (**B**), the reactivity to individual CD4 T cell epitopes, within HA, as measured by IL-2 EliSpot assays as spots per million CD4 T cells (See Table 1), were summed for each strain of mouse examined and placed into the "decades" of peptides spanning the entire sequence of the HA protein as described in results. All decades contain ten peptides except for the last set (peptides 91–94). Cumulative reactivity for each decade is represented as a cone with height corresponding to the summed spots per million (Y-axis) for each strain of mouse. Cones are colored for each MHC haplotype or MHC class II molecule for the various strains of mice utilized for this study as shown on the Z-axis.

```
MKAKLLVLLCTFTATYADTICIGYHANNSTDTVDTVLEKNVTVTHSVNLLEDSHNGKLCL  60

LKGIAPLQLGNCSVAGWILGNPECELLISKESWSYIVETPNPENGTCYPGYFADYEELRE  120

QLSSVSSFERFEIFPKESSWPNHTVTGVSASCSHNGKSSFYRNLLWLTGKNGLYPNLSK  179

SYVNNKEKEVLVLWGVHHPPNIGNQRALYHTENAYVSVVSSHYSRRFTPEIAKRPKVRDQ  239

EGRINYYWTLLEPGDTIIFEANGNLIAPWYAFALSRGFGSGIITSNAPMDECDAKCQTPQ  299

GAINSSLPFQNVHPVTIGECPKYVRSAKLRMVTGLRNIPSIQSRGLFGAIAGFIEGGWTG  359
                                            F
MVDGWYGYHHQNEQGSGYAADQKSTQNAINGITNKVNSVIEKMNTQFTAVGKEFNKLERR  419

MENLNKKVDDGFLDIWTYNAELLVLLENERTLDFHDSNVKNLYEKVKSQLKNNAKEIGNG  479

CFEFYHKCNNECMESVKNGTYDYPKYSEESKLNREKIDGVKLESMGVYQILAIYSTVASS  539
                                                          TM
LVLLVSLGAISFWMCSNGSLQCRICI  565
                     C
```

Zone	Amino Acids	
Dead Zones	1-119	MKAK-EELR
	232-315	KRPK-HPVT
	457-565	NVKN-RICI
Live Zones	120-190	EQLS-KEVL
	316-414	IGEC-KEFN
Infrequent/Minor	191-231	VLWG-PEIA
	415-456	KLER-FHDS
Fusion Peptide (F)	344-363	GLFG-MVDG
Transmembrane Domain (TM)	529-554	ILAI-WMCS
Cytosolic Tail (C)	555-565	NGSL-RICI
Cleavage site	343	

Figure 3. Stretches of the HA sequence that have distinct regions of CD4 T cell epitope dominance. The HA sequence is shown highlighted in regions that contain many dominant epitopes (shown in red), minor epitopes (shown in cream) or devoid of epitopes (shown highlighted in dark grey).

What becomes immediately clear from this rendition is the distinct structural segregation of the different zones of activity. The live regions, responsible for eliciting the majority of CD4 T cell epitopes, are located in regions that are highly structured, but also solvent exposed, suggesting that the stability of the tertiary structure coupled with accessibility to proteasomal processes is critical for the generation of CD4 T cell epitopes to HA. This feature is best illustrated by the alpha-helical domain in the stem region and the sheet structure in the head of the trimer. In contrast, the dead regions of HA are clustered in the membrane proximal region (Figure 4C), as well as in loosely structured regions between the head and stem (Figure 4A,B). The dead portion of HA found in the head of the trimer is in a well-ordered sheet structure, but is somewhat occluded by the overlaying red-colored sheet on the outside of the protein (Figure 4D), possibly inaccessible to proteases. The regions of HA designated as having minor reactivity are all relegated to the interior of the trimer, centralized in the triple-helical structure of the stem and the underlying base of the layered sheet structure of the head. It is likely that the regions of minor reactivity are under-represented as CD4 T cell epitopes due to both the highly ordered tertiary structure and the internal positioning within the HA protein.

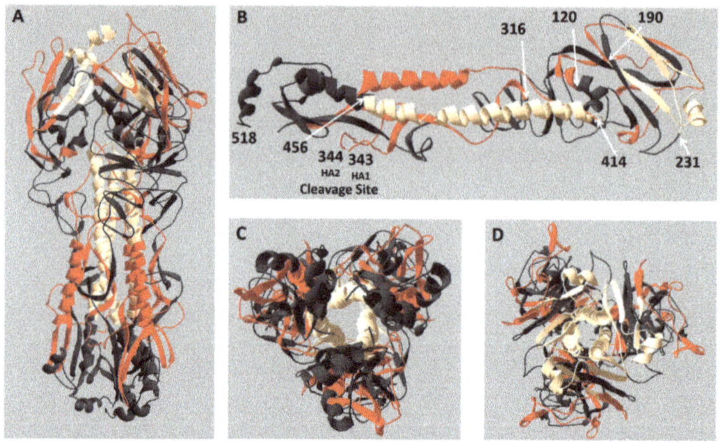

Figure 4. Localization of regions of HA that are enriched for or that lack CD4 T cell epitopes. Regions of CD4 T cell reactivity were mapped onto the trimeric structure of HA, with "Live" depicted in red, "Minor" in cream, and "Dead" in dark grey, categorized based on the data shown in Figure 2 are shown in (**A**). Amino acid residues at transition points between areas of differing reactivity are indicated on the HA monomer (**B**), as well as the HA cleavage site (amino acids 343, 344) and the carboxy terminus of the resolved structure (amino acid 518). The membrane proximal region (**C**, bottom view) of the trimer, as well as the head region (**D**, top view), are provided as an alternate vantage point of the HA protein structure. The structure is adapted from the H1N1 A/Brevig Mission/1/1918 hemagglutinin HA0 crystal structure (PDB: 1RD8), a close analog of A/New Caledonia/20/1999.

3. Discussion

The studies reported here suggest that in the primary response to influenza infection, only subsets of the segments of HA are selected by the elicited CD4 T cell repertoire. It is reasonable to speculate that such selection is multifactorial. Selection of CD4 T cell responses in the host has been shown to reflect at least in part the sequential processes in antigen presentation, including antigen uptake into endosomal compartments of antigen presenting cells (APC), pH-induced unfolding, reduction of disulfide bonds, proteolytic release of antigenic peptide, acquisition of the peptide by host MHC class II molecules and editing by HLA-DM and, finally the export of the peptide:class II complex to the cell surface of the APC (reviewed in [44–57]). Factors that have been implicated in previous studies for selection of dominant epitopes by CD4 T cells include proteolytic processing, three dimensional structure, sensitivity to proteolytic enzymes, or biochemical features of peptide:MHC class II complexes [56,58–63]. Selection of CD4 T cells is also dependent on the T cell receptor repertoire in the host that can recognize the presented peptide:class II complexes by APC [64–66].

It is interesting to consider the mechanisms that may underlie the characteristic dominant "hot spots" and "dead zones" of HA, representing regions that are enriched for CD4 T cell epitopes or that have a paucity of them, respectively. First, it is possible that the segments of HA within the hot spots of reactivity have atypically broad ability to bind to MHC class II molecules. However, because of the diversity of hosts that recognize these segments, each possessing distinct MHC class II molecules with their own binding preferences (reviewed in [67,68]), this seems less likely to be a dominant determining factor in selection of epitopes from these regions. As discussed, it is possible that the multimeric nature of the HA trimer limits proteolytic access of some internal sequestered regions of the trimeric HA, as is shown for the regions of the HA trimer that contain very few epitopes indicated in Figure 4. This feature of multimeric proteins may be particularly important in MHC class II epitope selection. The hot spots, shown in Figure 4, suggest that they tend to be on the solvent exposed regions that may be particularly accessible to proteolytic enzymes. Interestingly, the element of access to proteolytic sensitivity in viral proteins has been implicated in selection of CD4 T cell immunodominant regions of HIV gp140 protein [69] and Zika virus C and E proteins [70], as well as in studies designed to generally predict antigenic site for CD4 T cell recognition [45].

Another element that may be involved in epitope selection of HA is sites of N-linked glycosylation. Although it is known that some glycopeptides can elicit CD4 T cells (reviewed in [71,72]), it is possible that the sites of complex or very large N-linked glycosylation may shield segments of HA from proteolysis, as has been observed for the endosomal proteins LAMP-1 and LAMP-2 [73] and synthetic peptides [74]. Alternatively, the N-linked carbohydrate may obstruct CD4 T cell receptor recognition. Either of these events would diminish the frequency of CD4 T cells that can be elicited, even if these peptides can bind to MHC class II molecules [75]. We did note that the high epitopic regions of HA were devoid of potential N-linked glycosylation sites, except at the boundaries of the antigenic peptides (data not shown). Clearly, more experiments are needed to dissect the basis for the selection of restricted regions of HA in the primary response to influenza. However, it is interesting to consider the possibility that accumulation of sites that allow N-linked carbohydrates to be added, a well-documented phenomena in drifted Influenza A viruses, speculated to occur to shield HA from recognition of protective antibody recognition [76–78] may also serve to diminish CD4 T cell epitope generation.

An additional issue that is relevant to our studies relates to the nature of the priming antigen presenting cell after infection, the cellular mechanisms involved in class II:peptide epitope generation and the natural substrate for processing of HA and presentation of HA-derived peptides. Several recent publications [79–81] suggested that influenza infection might select for a distinct set of CD4 epitopes than those elicited by vaccination through a pathway initiated by infected antigen presenting cells. These data suggested to us that protein vaccination strategies, used extensively to provide protective immunity to influenza might misdirect the CD4 T cells response to a distinct subset of epitopes from those elicited by infection, diminishing the ability of influenza vaccines to CD4 T cells be recruited into the response to infection. To address this important issue, we recently we formally compared CD4 T cell epitope selection in mice via influenza infection vs. vaccination with recombinant NP and HA [35] revealed that the multitude of individual NP and HA epitopes elicited by these two modes of priming were the same. Moreover, we found that protein vaccination leads to memory CD4 T cells that can be recalled by infection. Thus, in the influenza model in mice that we have used in this study, as well as an earlier study from our group [12], suggest that the primary source of HA used for antigen presentation after infection likely involves intact HA released by infected or dying cells, rather than infectious virions.

In addition to considering the mechanisms that might underlie the CD4 T cell immunodominance of selected regions of HA, it is interesting to consider implications of these studies to human CD4 T cell immunity to influenza. Examining the CD4 T cell HA epitope specificity in the primary responses to influenza vaccination and infection is not possible in humans because of the extremely low frequency of individual-influenza CD4 T cells specific for single epitopes [14,16,20,82], coupled with the low blood volumes that it is possible to sample in infants and young children [83]. Our sampling of circulating adult human CD4 T cells specific for broad segments of HA suggest that regions of high genetic conservation are somewhat dominant in the circulating memory population [11,20,84]. It seems likely that epitope preferences within HA initiated during the first encounter of the human host might be shifted in specificity over time, due to the repeated and periodic confrontations of humans to alternative influenza strains generated by antigenic drift and shift. There are highly conserved regions of HA that will likely be repeatedly encountered in humans over a lifetime, particularly in the membrane proximal stem domain. HA-derived antigenic peptides derived from these regions may continually boost and expand CD4 T cells for these highly conserved regions, while specificities elicited in the primary response may decay in their relative frequency over time. It will be of interest to determine whether immunodominance hierarchies in HA shift toward these highly conserved regions as human subjects age or as shift and drift occurs, eliminating the potential of altered or eliminated CD4 T cell epitopes to boost the CD4 T cell repertoire upon influenza vaccination or infection.

4. Materials and Methods

4.1. Mice

The HLA-DR1 (B10.M/J-TgN-DR1) and HLA-DR4 (C57BL/6Tac-Abb<tm>TgNDR4) transgenic mice were obtained from D. Zaller (Merck) through Taconic Laboratories. The HLA-DQ8 (DQB.0302) transgenic mice were obtained from C. David [85]. B10.PL (B10.PL-H2^{u}H2-T18^{a}/(73NS)SnJ) mice were obtained from Jackson Laboratory. The DR1, DR4, DQ8, and B10.PL mice were bred and maintained in the specific pathogen free facility at the University of Rochester. Commercially available female C57BL/10, BALB/c, A/JCr, and SJL/JCr were purchased from National Cancer Institute-Frederick (Frederick, MD). All mice were used at 2–6 months of age.

4.2. Ethics Statement

All mice were maintained in the specific-pathogen free facility at the University of Rochester Medical Center according to the institutional guidelines. All animal protocols used in this study adhere to the AAALAC International, the Animal Welfare Act, and the PHS Guide and were approved by the University of Rochester Committee on Animal Resources, Animal Welfare Assurance Number A3291-01. The protocol under which these studies were conducted was originally approved 4 March 2006, (protocol no. 2006-030) and has been reviewed and re-approved every 36 months with the most recent review and approval 23 January 2018.

4.3. Peptides

17-mer peptides overlapping by 11 amino acids to encompass the entire sequence of the HA of influenza virus A/New Caledonia/20/99 (NR-2602) were obtained from BEI Resources, ATCC. The peptides were reconstituted at 10 mM in PBS, with or without added dimethyl sulfoxide, to increase solubility of hydrophobic peptides, and 1 mM dithiothreitol, for cysteine containing peptides. Working stocks (1 mM) were prepared in complete DMEM, (Dulbecco's Minimal Essential Media), sterilized and stored at −20 °C, as were concentrated stocks.

4.4. Infections, Tissue Harvest and Preparation of CD4 T Cell Enriched Cell Populations

A/New Caledonia/20/99 virus was produced as we have previously described [34,39]. Mice were infected intranasally at 20,000–100,000 EID_{50} in 30 ul of phosphate buffered saline (PBS) after being anesthetized by intraperitoneal injection with tribromoethanol (Avertin; 250–300 ul per mouse, normalized to weight). Spleens were excised at 10–13 days post-infection. For cytokine EliSpots, spleen cells were pooled from infected mice and used as the source of CD4 T cells. Cells were depleted of red blood cells (RBC) using ACK Lysis Buffer (0.15 M NH_4Cl, 1 mM $KHCO_3$, and 0.1 mM Na_2-EDTA in H_2O, pH 7.2). After washing, cells were depleted of B cells, CD8 cells, and class II positive cells using MACS negative selection (Miltenyi Biotec), according to the manufacturer instructions.

4.5. EliSpot Assays

EliSpot assays were performed as previously described [39]. Briefly, 96-well filter plates (Millipore) were coated with 2 ug/ml purified rat anti-mouse IL-2 (clone JES6-1A12, BD Biosciences) in PBS overnight at 4 °C, washed with media to remove unbound antibody and incubated with 100 ul media per well for 1 h to block non-specific binding. Isolated CD4 T cells were co-cultured with syngeneic spleen cells or transfected fibroblast antigen presenting cell (APC) (for HLA-DR1 transgenic mice [39]) and peptides for 16–20 h at 37 °C and 5% CO_2. The cells were removed from the plates, and the plates were washed with wash buffer (1X PBS, 0.1% Tween-20). Biotinylated rat anti-mouse IL-2 (clone JES6-5H4, BD Biosciences) was added at a concentration of 2 ug/ml, 50 ul/well, in wash buffer with 10% FBS (Fetal Bovine Serum) and incubated at room temperature for 30 min. The plates were washed again and streptavidin-conjugated alkaline phosphatase (Jackson Immuno Research) was added at a dilution of 1:1000 in wash buffer with 10% FBS, 50 ul/well, and incubated for 30 min at room temperature. The plates were and developed using Vector Blue substrate kit III (Vector Laboratories, City, CA, USA) prepared in 100 mM Tris, pH 8.2. After drying, quantification of spots was performed with an Immunospot reader series 5.2, using Immunospot software, version 5.1.

4.6. Structural Analyses

The structure of hemagglutinin (1RD8) [86] was analyzed for the internal positioning of CD4 T cell epitopes. This protein structure was acquired from the Protein Data Base (PDB ID indicated) and modified by color in Swiss-Pdbviewer to indicate by color the regions of HA that correspond to the abundance of CD4 T cell epitopes.

Supplementary Materials: The following are available online at http://www.mdpi.com/2076-0817/8/4/220/s1, Table S1: Peptides from H1N1 A/New Caledonia/20/99 and their corresponding amino acid sequence obtained from BEI, Table S2: Each mouse strain including those without epitopes >125 spots/million CD4 T cells. Those peptides in red are those included in Table 1, with the bold corresponding to those with ≥300 spots/million.

Author Contributions: A.J.S., Z.A.G.K., K.A.R., S.A.M. were involved in the methodology, data curation and formal analysis. A.J.S. wrote the original draft. K.A.R., Z.A.G.K., S.A.M. and A.J.S. were involved in review and editing of the manuscript.

Funding: This project has been funded in whole with Federal funds from the National Institute of Allergy and Infectious Diseases, National Institutes of Health, Department of Health and Human Services, under CEIRS Contract No. HHSN266200700008C and HHSN272201400005C to AJS. SM was supported by the Institutional Program Unifying Population and Laboratory Based Sciences Award from the Burroughs Wellcome Fund.

Conflicts of Interest: Authors declare no conflict of interest.

References

1. Krammer, F. The human antibody response to influenza A virus infection and vaccination. *Nat. Rev. Immunol.* **2019**. [CrossRef] [PubMed]
2. Dunkle, L.M.; Izikson, R. Recombinant hemagglutinin influenza vaccine provides broader spectrum protection. *Expert Rev. Vaccines* **2016**, *15*, 957–966. [CrossRef] [PubMed]
3. Treanor, J.J. Clinical Practice. Influenza Vaccination. *N. Engl. J. Med.* **2016**, *375*, 1261–1268. [CrossRef] [PubMed]
4. Reber, A.; Katz, J. Immunological assessment of influenza vaccines and immune correlates of protection. *Expert Rev. Vaccines* **2013**, *12*, 519–536. [CrossRef] [PubMed]
5. Nachbagauer, R.; Palese, P. Development of next generation hemagglutinin-based broadly protective influenza virus vaccines. *Curr. Opin. Immunol.* **2018**, *53*, 51–57. [CrossRef]
6. Padilla-Quirarte, H.O.; Lopez-Guerrero, D.V.; Gutierrez-Xicotencatl, L.; Esquivel-Guadarrama, F. Protective Antibodies Against Influenza Proteins. *Front. Immunol.* **2019**, *10*, 1677. [CrossRef]
7. Zhang, Y.; Xu, C.; Zhang, H.; Liu, G.D.; Xue, C.; Cao, Y. Targeting Hemagglutinin: Approaches for Broad Protection against the Influenza A Virus. *Viruses* **2019**, *11*, 405. [CrossRef] [PubMed]
8. Qi, H. T follicular helper cells in space-time. *Nat. Rev. Immunol.* **2016**, *16*, 612–625. [CrossRef]
9. Ueno, H. Tfh cell response in influenza vaccines in humans: What is visible and what is invisible. *Curr. Opin. Immunol.* **2019**, *59*, 9–14. [CrossRef]
10. Crotty, S. A brief history of T cell help to B cells. *Nat. Rev. Immunol.* **2015**, *15*, 185–189. [CrossRef]
11. Nayak, J.L.; Richards, K.A.; Yang, H.; Treanor, J.J.; Sant, A.J. Effect of influenza A(H5N1) vaccine prepandemic priming on CD4+ T-cell responses. *J. Infect. Dis.* **2015**, *211*, 1408–1417. [CrossRef] [PubMed]

12. Alam, S.; Knowlden, Z.A.; Sangster, M.Y.; Sant, A.J. CD4 T cell help is limiting and selective during the primary B cell response to influenza virus infection. *J. Virol.* **2014**, *88*, 314–324. [CrossRef] [PubMed]
13. Savic, M.; Dembinski, J.L.; Kim, Y.; Tunheim, G.; Cox, R.J.; Oftung, F.; Peters, B.; Mjaaland, S. Epitope specific T-cell responses against influenza A in a healthy population. *Immunology* **2016**, *147*, 165–177. [CrossRef] [PubMed]
14. Uchtenhagen, H.; Rims, C.; Blahnik, G.; Chow, I.T.; Kwok, W.W.; Buckner, J.H.; James, E.A. Efficient ex vivo analysis of CD4+ T-cell responses using combinatorial HLA class II tetramer staining. *Nat. Commun.* **2016**, *7*, 12614. [CrossRef]
15. Yang, J.; James, E.; Gates, T.J.; DeLong, J.H.; LaFond, R.E.; Malhotra, U.; Kwok, W.W. CD4+ T cells recognize unique and conserved 2009 H1N1 influenza hemagglutinin epitopes after natural infection and vaccination. *Int. Immunol.* **2013**, *25*, 447–457. [CrossRef]
16. Roti, M.; Yang, J.; Berger, D.; Huston, L.; James, E.A.; Kwok, W.W. Healthy human subjects have CD4+ T cells directed against H5N1 influenza virus. *J. Immunol.* **2008**, *180*, 1758–1768. [CrossRef]
17. Yang, J.; James, E.A.; Huston, L.; Danke, N.A.; Liu, A.W.; Kwok, W.W. Multiplex mapping of CD4 T cell epitopes using class II tetramers. *Clin. Immunol.* **2006**, *120*, 21–32. [CrossRef]
18. Richards, K.A.; Treanor, J.J.; Nayak, J.L.; Sant, A.J. Overarching Immunodominance Patterns and Substantial Diversity in Specificity and Functionality in the Circulating Human Influenza A and B CD4 T Cell Repertoire. *J. Infect. Dis.* **2018**. [CrossRef]
19. Leddon, S.A.; Richards, K.A.; Treanor, J.J.; Sant, A.J. Abundance and specificity of influenza reactive circulating memory follicular helper and non-follicular helper CD4 T cells in healthy adults. *Immunology* **2015**, *146*, 157–162. [CrossRef]
20. Richards, K.A.; Nayak, J.; Chaves, F.A.; DiPiazza, A.; Knowlden, Z.A.; Alam, S.; Treanor, J.J.; Sant, A.J. Seasonal Influenza Can Poise Hosts for CD4 T-Cell Immunity to H7N9 Avian Influenza. *J. Infect. Dis.* **2015**, *212*, 86–94. [CrossRef]
21. Richards, K.A.; Topham, D.; Chaves, F.A.; Sant, A.J. Cutting edge: CD4 T cells generated from encounter with seasonal influenza viruses and vaccines have broad protein specificity and can directly recognize naturally generated epitopes derived from the live pandemic H1N1 virus. *J. Immunol.* **2010**, *185*, 4998–5002. [CrossRef] [PubMed]
22. Sant, A.J.; DiPiazza, A.T.; Nayak, J.L.; Rattan, A.; Richards, K.A. CD4 T cells in protection from influenza virus: Viral antigen specificity and functional potential. *Immunol. Rev.* **2018**, *284*, 91–105. [CrossRef] [PubMed]
23. Sant, A.J.; Richards, K.A.; Nayak, J. Distinct and complementary roles of CD4 T cells in protective immunity to influenza virus. *Curr. Opin. Immunol.* **2018**, *53*, 13–21. [CrossRef] [PubMed]
24. Babon, J.A.; Cruz, J.; Ennis, F.A.; Yin, L.; Terajima, M. A human CD4+ T cell epitope in the influenza hemagglutinin is cross-reactive to influenza A virus subtypes and to influenza B virus. *J. Virol.* **2012**, *86*, 9233–9243. [CrossRef] [PubMed]
25. Moise, L.; Tassone, R.; Latimer, H.; Terry, F.; Levitz, L.; Haran, J.P.; Ross, T.M.; Boyle, C.M.; Martin, W.D.; De Groot, A.S. Immunization with cross-conserved H1N1 influenza CD4+ T-cell epitopes lowers viral burden in HLA DR3 transgenic mice. *Hum. Vaccin. Immunother.* **2013**, *9*, 2060–2068. [CrossRef]
26. Alam, S.; Sant, A.J. Infection with seasonal influenza virus elicits CD4 T cells specific for genetically conserved epitopes that can be rapidly mobilized for protective immunity to pandemic H1N1 influenza virus. *J. Virol.* **2011**, *85*, 13310–13321. [CrossRef]
27. Ye, M.; Kasey, S.; Khurana, S.; Nguyen, N.T.; Schubert, S.; Nugent, C.T.; Kuus-Reichel, K.; Hampl, J. MHC class II tetramers containing influenza hemagglutinin and EBV EBNA1 epitopes detect reliably specific CD4(+) T cells in healthy volunteers. *Hum. Immunol.* **2004**, *65*, 507–513. [CrossRef]
28. Wang, S.F.; Yao, L.; Liu, S.J.; Chong, P.; Liu, W.T.; Chen, Y.M.; Huang, J.C. Identifying conserved DR1501-restricted CD4(+) T-cell epitopes in avian H5N1 hemagglutinin proteins. *Viral. Immunol.* **2010**, *23*, 585–593. [CrossRef]
29. Duvvuri, V.R.; Moghadas, S.M.; Guo, H.; Duvvuri, B.; Heffernan, J.M.; Fisman, D.N.; Wu, G.E.; Wu, J. Highly conserved cross-reactive CD4+ T-cell HA-epitopes of seasonal and the 2009 pandemic influenza viruses. *Influenza Respir. Viruses* **2010**, *4*, 249–258. [CrossRef]
30. Fehlings, M.; Chakarov, S.; Simoni, Y.; Sivasankar, B.; Ginhoux, F.; Newell, E.W. Multiplex peptide-MHC tetramer staining using mass cytometry for deep analysis of the influenza-specific T-cell response in mice. *J. Immunol. Methods* **2018**, *453*, 30–36. [CrossRef]

31. Herati, R.S.; Muselman, A.; Vella, L.; Bengsch, B.; Parkhouse, K.; Del Alcazar, D.; Kotzin, J.; Doyle, S.A.; Tebas, P.; Hensley, S.E.; et al. Successive annual influenza vaccination induces a recurrent oligoclonotypic memory response in circulating T follicular helper cells. *Sci. Immunol.* **2017**, *2*. [CrossRef] [PubMed]
32. Lucas, M.; Day, C.L.; Wyer, J.R.; Cunliffe, S.L.; Loughry, A.; McMichael, A.J.; Klenerman, P. Ex vivo phenotype and frequency of influenza virus-specific CD4 memory T cells. *J. Virol.* **2004**, *78*, 7284–7287. [CrossRef] [PubMed]
33. Nayak, J.L.; Sant, A.J. Loss in CD4 T-cell responses to multiple epitopes in influenza due to expression of one additional MHC class II molecule in the host. *Immunology* **2012**, *136*, 425–436. [CrossRef]
34. Nayak, J.L.; Richards, K.A.; Chaves, F.A.; Sant, A.J. Analyses of the specificity of CD4 T cells during the primary immune response to influenza virus reveals dramatic MHC-linked asymmetries in reactivity to individual viral proteins. *Viral. Immunol.* **2010**, *23*, 169–180. [CrossRef]
35. Rattan, A.; Richards, K.A.; Knowlden, Z.A.G.; Sant, A.J. Protein vaccination directs the CD4+ T cell response towards shared protective epitopes that can be recalled after influenza infection. *J. Virol.* **2019**. [CrossRef]
36. Richards, K.A.; DiPiazza, A.T.; Rattan, A.; Knowlden, Z.A.G.; Yang, H.; Sant, A.J. Diverse Epitope Specificity, Immunodominance Hierarchy, and Functional Avidity of Effector CD4 T Cells Established During Priming Is Maintained in Lung After Influenza A Virus Infection. *Front. Immunol.* **2018**, *9*, 655. [CrossRef]
37. DiPiazza, A.; Laniewski, N.; Rattan, A.; Topham, D.J.; Miller, J.; Sant, A.J. CD4 T Cell Epitope Specificity and Cytokine Potential Are Preserved as Cells Transition from the Lung Vasculature to Lung Tissue following Influenza Virus Infection. *J. Virol.* **2018**, *92*. [CrossRef]
38. Knowlden, Z.A.; Sant, A.J. CD4 T cell epitope specificity determines follicular versus non-follicular helper differentiation in the polyclonal response to influenza infection or vaccination. *Sci. Rep.* **2016**, *6*, 28287. [CrossRef]
39. Richards, K.A.; Chaves, F.A.; Krafcik, F.R.; Topham, D.J.; Lazarski, C.A.; Sant, A.J. Direct ex vivo analyses of HLA-DR1 transgenic mice reveal an exceptionally broad pattern of immunodominance in the primary HLA-DR1-restricted CD4 T-cell response to influenza virus hemagglutinin. *J. Virol.* **2007**, *81*, 7608–7619. [CrossRef]
40. Richards, K.A.; Chaves, F.A.; Sant, A.J. Infection of HLA-DR1 transgenic mice with a human isolate of influenza a virus (H1N1) primes a diverse CD4 T-cell repertoire that includes CD4 T cells with heterosubtypic cross-reactivity to avian (H5N1) influenza virus. *J. Virol.* **2009**, *83*, 6566–6577. [CrossRef]
41. Sundararajan, A.; Huan, L.; Richards, K.A.; Marcelin, G.; Alam, S.; Joo, H.; Yang, H.; Webby, R.J.; Topham, D.J.; Sant, A.J.; et al. Host differences in influenza-specific CD4 T cell and B cell responses are modulated by viral strain and route of immunization. *PLoS ONE* **2012**, *7*, e34377. [CrossRef] [PubMed]
42. Tobery, T.W.; Wang, S.; Wang, X.M.; Neeper, M.P.; Jansen, K.U.; McClements, W.L.; Caulfield, M.J. A simple and efficient method for the monitoring of antigen-specific T cell responses using peptide pool arrays in a modified ELISpot assay. *J. Immunol. Methods* **2001**, *254*, 59–66. [CrossRef]
43. Chaves, F.A.; Lee, A.H.; Nayak, J.L.; Richards, K.A.; Sant, A.J. The utility and limitations of current Web-available algorithms to predict peptides recognized by CD4 T cells in response to pathogen infection. *J. Immunol.* **2012**, *188*, 4235–4248. [CrossRef] [PubMed]
44. Schneidman-Duhovny, D.; Khuri, N.; Dong, G.Q.; Winter, M.B.; Shifrut, E.; Friedman, N.; Craik, C.S.; Pratt, K.P.; Paz, P.; Aswad, F.; et al. Predicting CD4 T-cell epitopes based on antigen cleavage, MHCII presentation, and TCR recognition. *PLoS ONE* **2018**, *13*, e0206654. [CrossRef] [PubMed]
45. Mettu, R.R.; Charles, T.; Landry, S.J. CD4+ T-cell epitope prediction using antigen processing constraints. *J. Immunol. Methods* **2016**, *432*, 72–81. [CrossRef] [PubMed]
46. Kim, A.; Ishizuka, I.E.; Hartman, I.Z.; Poluektov, Y.; Narayan, K.; Sadegh-Nasseri, S. Studying MHC Class II Peptide Loading and Editing In Vitro. *Methods Mol. Biol.* **2019**, *1988*, 343–355. [CrossRef] [PubMed]
47. Jurewicz, M.M.; Stern, L.J. Class II MHC antigen processing in immune tolerance and inflammation. *Immunogenetics* **2019**, *71*, 171–187. [CrossRef]
48. Kelly, A.; Trowsdale, J. Genetics of antigen processing and presentation. *Immunogenetics* **2019**, *71*, 161–170. [CrossRef]
49. Thomas, C.; Tampe, R. Proofreading of Peptide-MHC Complexes through Dynamic Multivalent Interactions. *Front. Immunol.* **2017**, *8*, 65. [CrossRef]
50. Rock, K.L.; Reits, E.; Neefjes, J. Present Yourself! By MHC Class I and MHC Class II Molecules. *Trends Immunol.* **2016**, *37*, 724–737. [CrossRef]

51. Roche, P.A.; Cresswell, P. Antigen Processing and Presentation Mechanisms in Myeloid Cells. *Microbiol. Spectr.* **2016**, *4*. [CrossRef]
52. Unanue, E.R.; Turk, V.; Neefjes, J. Variations in MHC Class II Antigen Processing and Presentation in Health and Disease. *Annu. Rev. Immunol.* **2016**, *34*, 265–297. [CrossRef] [PubMed]
53. Roche, P.A.; Furuta, K. The ins and outs of MHC class II-mediated antigen processing and presentation. *Nat. Rev. Immunol.* **2015**, *15*, 203–216. [CrossRef]
54. Mintern, J.D.; Macri, C.; Villadangos, J.A. Modulation of antigen presentation by intracellular trafficking. *Curr. Opin. Immunol.* **2015**, *34*, 16–21. [CrossRef]
55. Rossjohn, J.; Gras, S.; Miles, J.J.; Turner, S.J.; Godfrey, D.I.; McCluskey, J. T cell antigen receptor recognition of antigen-presenting molecules. *Ann. Rev. Immunol.* **2015**, *33*, 169–200. [CrossRef]
56. Sant, A.J.; Chaves, F.A.; Jenks, S.A.; Richards, K.A.; Menges, P.; Weaver, J.M.; Lazarski, C.A. The relationship between immunodominance, DM editing, and the kinetic stability of MHC class II:peptide complexes. *Immunol. Rev.* **2005**, *207*, 261–278. [CrossRef]
57. Kambayashi, T.; Laufer, T.M. Atypical MHC class II-expressing antigen-presenting cells: Can anything replace a dendritic cell? *Nat. Rev. Immunol.* **2014**, *14*, 719–730. [CrossRef]
58. Kim, A.; Sadegh-Nasseri, S. Determinants of immunodominance for CD4 T cells. *Curr. Opin. Immunol.* **2015**, *34*, 9–15. [CrossRef]
59. Hastings, K.T. GILT: Shaping the MHC Class II-Restricted Peptidome and CD4(+) T Cell-Mediated Immunity. *Front. Immunol.* **2013**, *4*, 429. [CrossRef]
60. Haque, A.; Blum, J.S. New insights in antigen processing and epitope selection: Development of novel immunotherapeutic strategies for cancer, autoimmunity and infectious diseases. *J. Biol. Regul. Homeost. Agents* **2005**, *19*, 93–104.
61. Mellins, E.D.; Stern, L.J. HLA-DM and HLA-DO, key regulators of MHC-II processing and presentation. *Curr. Opin. Immunol.* **2014**, *26*, 115–122. [CrossRef]
62. Pos, W.; Sethi, D.K.; Wucherpfennig, K.W. Mechanisms of peptide repertoire selection by HLA-DM. *Trends Immunol* **2013**, *34*, 495–501. [CrossRef] [PubMed]
63. Sant, A.J.; Chaves, F.A.; Leddon, S.A.; Tung, J. The control of the specificity of CD4 T cell responses: Thresholds, breakpoints, and ceilings. *Front. Immunol.* **2013**, *4*, 340. [CrossRef] [PubMed]
64. Hassler, T.; Urmann, E.; Teschner, S.; Federle, C.; Dileepan, T.; Schober, K.; Jenkins, M.K.; Busch, D.H.; Hinterberger, M.; Klein, L. Inventories of naive and tolerant mouse CD4 T cell repertoires reveal a hierarchy of deleted and diverted T cell receptors. *Proc. Natl. Acad. Sci. USA* **2019**, *116*, 18537–18543. [CrossRef] [PubMed]
65. Jenkins, M.K.; Moon, J.J. The role of naive T cell precursor frequency and recruitment in dictating immune response magnitude. *J. Immunol* **2012**, *188*, 4135–4140. [CrossRef] [PubMed]
66. Zarnitsyna, V.I.; Evavold, B.D.; Schoettle, L.N.; Blattman, J.N.; Antia, R. Estimating the diversity, completeness, and cross-reactivity of the T cell repertoire. *Front. Immunol.* **2013**, *4*, 485. [CrossRef]
67. Nielsen, M.; Lund, O.; Buus, S.; Lundegaard, C. MHC class II epitope predictive algorithms. *Immunology* **2010**, *130*, 319–328. [CrossRef]
68. Vaughan, K.; Xu, X.; Caron, E.; Peters, B.; Sette, A. Deciphering the MHC-associated peptidome: A review of naturally processed ligand data. *Expert Rev. Proteomics* **2017**, *14*, 729–736. [CrossRef]
69. Surman, S.; Lockey, T.D.; Slobod, K.S.; Jones, B.; Riberdy, J.M.; White, S.W.; Doherty, P.C.; Hurwitz, J.L. Localization of CD4+ T cell epitope hotspots to exposed strands of HIV envelope glycoprotein suggests structural influences on antigen processing. *Proc. Natl. Acad. Sci. USA* **2001**, *98*, 4587–4592. [CrossRef]
70. Koblischke, M.; Stiasny, K.; Aberle, S.W.; Malafa, S.; Tsouchnikas, G.; Schwaiger, J.; Kundi, M.; Heinz, F.X.; Aberle, J.H. Structural Influence on the Dominance of Virus-Specific CD4 T Cell Epitopes in Zika Virus Infection. *Front. Immunol* **2018**, *9*, 1196. [CrossRef]
71. Avci, F.Y.; Li, X.; Tsuji, M.; Kasper, D.L. Carbohydrates and T cells: A sweet twosome. *Semin. Immunol.* **2013**, *25*, 146–151. [CrossRef] [PubMed]
72. Sun, L.; Middleton, D.R.; Wantuch, P.L.; Ozdilek, A.; Avci, F.Y. Carbohydrates as T-cell antigens with implications in health and disease. *Glycobiology* **2016**, *26*, 1029–1040. [CrossRef] [PubMed]
73. Kundra, R.; Kornfeld, S. Asparagine-linked oligosaccharides protect Lamp-1 and Lamp-2 from intracellular proteolysis. *J. Biol. Chem.* **1999**, *274*, 31039–31046. [CrossRef] [PubMed]

74. Hung, M.E.; Leonard, J.N. Stabilization of exosome-targeting peptides via engineered glycosylation. *J. Biol. Chem.* **2015**, *290*, 8166–8172. [CrossRef] [PubMed]
75. Galli-Stampino, L.; Meinjohanns, E.; Frische, K.; Meldal, M.; Jensen, T.; Werdelin, O.; Mouritsen, S. T-cell recognition of tumor-associated carbohydrates: The nature of the glycan moiety plays a decisive role in determining glycopeptide immunogenicity. *Cancer Res.* **1997**, *57*, 3214–3222. [PubMed]
76. Altman, M.O.; Angel, M.; Kosik, I.; Trovao, N.S.; Zost, S.J.; Gibbs, J.S.; Casalino, L.; Amaro, R.E.; Hensley, S.E.; Nelson, M.I.; et al. Human Influenza A Virus Hemagglutinin Glycan Evolution Follows a Temporal Pattern to a Glycan Limit. *MBio* **2019**, *10*. [CrossRef] [PubMed]
77. Das, S.R.; Puigbo, P.; Hensley, S.E.; Hurt, D.E.; Bennink, J.R.; Yewdell, J.W. Glycosylation focuses sequence variation in the influenza A virus H1 hemagglutinin globular domain. *PLoS Pathog.* **2010**, *6*, e1001211. [CrossRef]
78. Tsuchiya, E.; Sugawara, K.; Hongo, S.; Matsuzaki, Y.; Muraki, Y.; Li, Z.N.; Nakamura, K. Effect of addition of new oligosaccharide chains to the globular head of influenza A/H2N2 virus haemagglutinin on the intracellular transport and biological activities of the molecule. *J. Gen. Virol.* **2002**, *83*, 1137–1146. [CrossRef]
79. Miller, M.A.; Ganesan, A.P.; Luckashenak, N.; Mendonca, M.; Eisenlohr, L.C. Endogenous antigen processing drives the primary CD4+ T cell response to influenza. *Nat. Med.* **2015**, *21*, 1216–1222. [CrossRef]
80. Eisenlohr, L.C.; Luckashenak, N.; Apcher, S.; Miller, M.A.; Sinnathamby, G. Beyond the classical: Influenza virus and the elucidation of alternative MHC class II-restricted antigen processing pathways. *Immunol. Res.* **2011**, *51*, 237–248. [CrossRef]
81. Mintern, J.D.; Villadangos, J.A. Antigen-presenting cells look within during influenza infection. *Nat. Med.* **2015**, *21*, 1123–1125. [CrossRef] [PubMed]
82. Danke, N.A.; Kwok, W.W. HLA class II-restricted CD4+ T cell responses directed against influenza viral antigens postinfluenza vaccination. *J. Immunol.* **2003**, *171*, 3163–3169. [CrossRef] [PubMed]
83. Howie, S.R. Blood sample volumes in child health research: Review of safe limits. *Bull. World Health Organ.* **2011**, *89*, 46–53. [CrossRef]
84. Nayak, J.L.; Fitzgerald, T.F.; Richards, K.A.; Yang, H.; Treanor, J.J.; Sant, A.J. CD4+ T-cell expansion predicts neutralizing antibody responses to monovalent, inactivated 2009 pandemic influenza A(H1N1) virus subtype H1N1 vaccine. *J. Infect. Dis* **2013**, *207*, 297–305. [CrossRef] [PubMed]
85. Cheng, S.; Smart, M.; Hanson, J.; David, C.S. Characterization of HLA DR2 and DQ8 transgenic mouse with a new engineered mouse class II deletion, which lacks all endogenous class II genes. *J. Autoimmun.* **2003**, *21*, 195–199. [CrossRef]
86. Stevens, J.; Corper, A.L.; Basler, C.F.; Taubenberger, J.K.; Palese, P.; Wilson, I.A. Structure of the uncleaved human H1 hemagglutinin from the extinct 1918 influenza virus. *Science* **2004**, *303*, 1866–1870. [CrossRef]

pathogens

MDPI

Article

Increasing the Safety Profile of the Master Donor Live Attenuated Influenza Vaccine

Thomas A. Hilimire [1,†], Aitor Nogales [1,2,†], Kevin Chiem [1,‡], Javier Ortego [2] and Luis Martinez-Sobrido [1,*,‡]

1 Department of Microbiology and Immunology, University of Rochester Medical Center, 601 Elmwood Avenue, Rochester, NY 14642, USA; Thomas_hilimire@urmc.rochester.edu (T.A.H.); nogales.aitor@inia.es (A.N.); Kevin_chiem@urmc.rochester.edu (K.C.)

2 Center for Animal Health Research, INIA-CISA, 28130 Valdeolmos, Madrid, Spain; ortego@inia.es

* Correspondence: l.martinez@txbiomed.org; Tel.: +1-585-259-8739

† These authors contributed equally to this work.

‡ Current address: Texas Biomedical Research Institute, 8715 W. Military Dr., San Antonio, TX 78212, USA.

Received: 16 September 2019; Accepted: 22 January 2020; Published: 29 January 2020

Abstract: Seasonal influenza epidemics remain one of the largest public health burdens nowadays. The best and most effective strategy to date in preventing influenza infection is a worldwide vaccination campaign. Currently, two vaccines are available to the public for the treatment of influenza infection, the chemically Inactivated Influenza Vaccine (IIV) and the Live Attenuated Influenza Vaccine (LAIV). However, the LAIV is not recommended for parts of the population, such as children under the age of two, immunocompromised individuals, the elderly, and pregnant adults. In order to improve the safety of the LAIV and make it available to more of the population, we sought to further attenuate the LAIV. In this study, we demonstrate that the influenza A virus (IAV) master donor virus (MDV) A/Ann Arbor/6/60 H2N2 LAIV can inhibit host gene expression using both the PA-X and NS1 proteins. Furthermore, we show that by removing PA-X, we can limit the replication of the MDV LAIV in a mouse model, while maintaining full protective efficacy. This work demonstrates a broadly applicable strategy of tuning the amount of host antiviral responses induced by the IAV MDV for the development of newer and safer LAIVs. Moreover, our results also demonstrate, for the first time, the feasibility of genetically manipulating the backbone of the IAV MDV to improve the efficacy of the current IAV LAIV.

Keywords: live attenuated influenza vaccine; influenza; master donor virus; virus–host interaction; vaccine safety; immunogenicity; protection efficacy

1. Introduction

The influenza A virus (IAV) belongs to the Orthomyxoviridae family of enveloped viruses, which contain an eight-segmented, negative-sense, single-stranded RNA genome [1]. According to the antigenic properties of the hemagglutinin (HA; H1–H18) and neuraminidase (NA; N1–N11) viral surface glycoproteins, IAVs are classified in subtypes with 18 HA and 11 NA subtypes currently circulating [2]. In humans, IAV is responsible for seasonal epidemics that cause mild to severe respiratory illness or even death, representing a threat worldwide to public health. Vaccination is considered the most effective approach to protect against seasonal influenza infections. However, despite worldwide vaccination plans, seasonal influenza is still responsible for 1 billion infections, causing 3–5 million cases of severe disease, and between 290,000 and 650,000 deaths annually, according to the World Health Organization (WHO) and the Center for Disease Control (CDC) [3]. Currently, seasonal IAVs circulating in humans include the H3N2 and the H1N1 subtypes [4]. However due to constant antigenic drift, vaccines have to be reformulated yearly to ensure that the HA and NA proteins included in the vaccine match those present in the seasonally circulating virus.

Inactivated influenza vaccines (IIVs), which induce limited T cell-mediated immune responses [5], have limited protection efficacy when the seasonal vaccine does not match the circulating strains [6,7]. On the other hand, live attenuated influenza vaccines (LAIVs) usually induce more robust adaptive B- and T-cell immune responses associated with better protection against homologous and heterologous influenza viruses [5–7]. Despite the multiple advantages of LAIVs, the current LAIV remains restricted for use in healthy children older than two and non-pregnant adults [8]. Therefore, there is a need to develop safer LAIVs that can be used more broadly in the human population, while maintaining their immunogenic properties. Current LAIVs for the treatment of human IAV infections contain the internal genes (*PB2, PB1, PA, NP, M,* and *NS*) of a master donor virus (MDV) and the surface HA and NA segments from the seasonal virus strains (H1N1 and H3N2) [9]. The MDV in use in the United States (US) is the A/Ann Arbor/6/60 H2N2 (A/AA/6/60), which was generated after several passages under suboptimal low temperatures [10]. The MDV A/AA/6/60 LAIV has a temperature sensitive (*ts*), cold adapted (*ca*), and attenuated (*att*) phenotype associated with the ability of the virus to replicate in the upper respiratory track (lower temperature) but not in the lower respiratory tract (higher temperatures) [11]. The *ca, ts,* and *att* signature of the MDV A/AA/6/60 LAIV is conferred by five mutations in three internal viral genes: the polymerase basic 2 (PB2; N265S) and 1 (PB1; K391E, D581G, and A661T) proteins and the viral nucleoprotein (NP; D34G) [10,12].

IAV has developed several mechanisms to counteract host antiviral responses, namely inhibiting the production of interferon (IFN) and the downstream activities of IFN-stimulated gene (ISG) proteins, which normally inhibit virus replication and propagation [13,14]. Segment 3 (PA) of IAV encodes two proteins, the first being the polymerase acid (PA) protein that is produced directly from the PA mRNA and has endonuclease activity, as well as being a component, together with PB2 and PB1, of the viral polymerase complex [15]. Segment 3 also encodes a second protein, PA-X, which is translated from a +1 frameshift open reading frame (ORF) located in the PA viral segment. PA-X shares the first N-terminal 191 amino acids with PA, but contains a unique short C-terminal sequence [15–18]. Importantly, PA-X has been shown to have multiple functions, such as the selective degradation of host RNA polymerase II-transcribed mRNAs, which leads to the selective inhibition of cellular protein synthesis, blocking of antiviral responses, or modulating host inflammation [15,19–23]. Despite PA-X and PA sharing the same N-terminal region, PA-X has a stronger endonucleolytic activity, indicating that the C-terminal domain is responsible for the cellular shutoff [24]. Furthermore, the primary transcript produced from the viral genome segment 8 (NS) of IAV is the non-structural protein 1 (NS1), a multifunctional protein which counteracts the innate immune responses, allowing the virus to replicate in IFN-competent systems [14,25–28]. Synergistically to PA-X, the NS1 protein of certain IAV strains can inhibit host protein synthesis, controlling the expression of IFN and/or ISGs [14,27,29–31]. To accomplish this, NS1 binds to the 30 kDa subunit of the cleavage and polyadenylation specificity factor (CPSF30), inhibiting the recognition by the CPSF complex of polyadenylation signals of mRNAs during transcription, blocking the cleavage of immature mRNAs and the addition of the poly (A) tail; this is because this poly (A) tail is required for nucleus export, stability, and translation of cellular mRNAs. The unprocessed mRNAs accumulate in the nucleus, leading to an inhibition of host gene expression, including IFN or ISGs [26,32–34]. The amino acid residues responsible for this NS1 function have been mapped in multiple IAV strains. For instance, A/Puerto Rico/8/34 H1N1 cannot bind CPSF30 due to mutations at positions 103 and 106, but that capacity can be restored by introducing amino acid changes at these residues (L103F and I106M) [30,35].

We have postulated that the ability of IAV NS1 and/or PA-X to inhibit innate immune responses might be modulated to generate more effective and/or safer LAIV approaches. In fact, we have generated LAIV-encoding PA-X and NS1 proteins with different abilities to inhibit host gene expression, using the backbone of an A/California/04/09 pandemic (p)H1N1 LAIV, demonstrating the feasibility of implementing this approach, alone or in combination with other methodologies, for the development of a novel LAIVs [18]. Here, we have evaluated whether the current MDV A/AA/6/60 LAIV used for the preparation of the seasonal human LAIV could be improved either in safety and/or immunogenicity by modulating the ability of NS1 and/or PA-X proteins to block host gene expression. To this end, first we evaluated if the NS1 and PA-X proteins of the MDV A/AA/6/60 LAIV have the ability to inhibit host gene expression. Next, we engineered a set of MDV A/AA/6/60 LAIVs encoding PA-X and NS1 proteins with different abilities to inhibit host gene expression, alone or in combination, and the HA and NA of pH1N1. The replication capability of these viruses was evaluated in vitro and their safety, immunogenicity, and protection efficacy against a homologous pH1N1 challenge were assessed in vivo using a mouse model of IAV infection. Our studies indicate that removing PA-X expression by altering the frameshift site, the MDV A/AA/6/60 LAIV can be further attenuated while still retaining a similar immunogenic and protective profile. The implementation of this approach could help to develop safer LAIVs than the currently commercially available IAV LAIV.

2. Results

2.1. The MDV A/AA/6/60 LAIV PA-X and NS1 Proteins Inhibit Host Gene Expression

Recently, we have shown that IAV PA-X and NS1 proteins can affect viral pathogenicity and the safety and immunogenicity of a pH1N1 LAIV candidate [18]. Therefore, we wanted to examine how PA-X and NS1 of the MDV A/AA/6/60 LAIV affected host gene expression (Figure 1). To this end, we used a well-established approach to evaluate the ability of NS1 and PA-X proteins to block host gene expression [26,27,35–37]. Human 293T cells were co-transfected with pCAGGS plasmids expressing the green fluorescent protein (GFP) or Gaussia luciferase (Gluc), along with pDZ plasmids that were either empty (E), contained the WT PA from the MDV A/AA/6/60 LAIV (PA_{WT}^{+}), or a mutant PA that has an altered frameshift (PA_{MT}^{-}). At 24 h post-transfection (p.t.), cells were imaged under a fluorescent microscope for GFP expression while tissue culture supernatants (TCS) were collected to quantify Gluc using a luminometer. The PA-X of the MDV A/AA/6/60 LAIV had a strong inhibitory activity showing almost complete knock down of GFP (Figure 1A) and a large decrease in Gluc (Figure 1B) expression. In comparison, the mutant *PA* gene that contained an altered frameshift motif, which is unable to express PA-X, showed no inhibition of GFP or Gluc protein synthesis (Figure 1A,B, respectively). Together these results indicate that the WT PA from the MDV A/AA/6/60 LAIV expresses a functional PA-X that inhibits host gene expression, and that disruption of the frameshift prevents the inhibitory effect mediated by PA-X. To verify that the PA protein was being expressed we also performed Western blot analysis (Figure 1C). We were able to detect similar expression levels of both WT and mutant PA proteins (Figure 1C). We were unable to detect directly PA-X expression in these assays, as the frameshift efficiency is only 1.3% for the WT *PA* gene [15]. Next, we examined the ability of MDV A/AA/6/60 LAIV NS1 to inhibit protein synthesis. Similar to PA-X, we co-transfected the two plasmids expressing the GFP and Gluc reporter genes with either empty (E), WT NS1 ($NS1_{WT}^{+}$), or mutant NS1 ($NS1_{MT}^{-}$) pDZ plasmids. The WT MDV A/AA/6/60 LAIV NS1 protein was able to effectively inhibit expression of both GFP (Figure 1D) and Gluc (Figure 1E) reporter proteins while the NS1 containing the mutations F103S and M106I in the CPSF30 binding domain showed no inhibitory effect in GFP or Gluc expression (Figure 1D,E, respectively). We also analyzed NS1 protein expression by Western blot (Figure 1F). While both WT and mutant NS1 proteins were expressed from their respective plasmids, the mutant NS1 showed higher expression levels as compared to the WT NS1, because of the ability of WT NS1 to inhibit its own synthesis (Figure 1F) [38]. These results demonstrate that WT NS1 from the MDV A/AA/6/60 LAIV inhibits protein expression and that mutations F103S and M106I, previously described to affect binding to CPSF30 [35], remove its inhibitory effect on host gene expression.

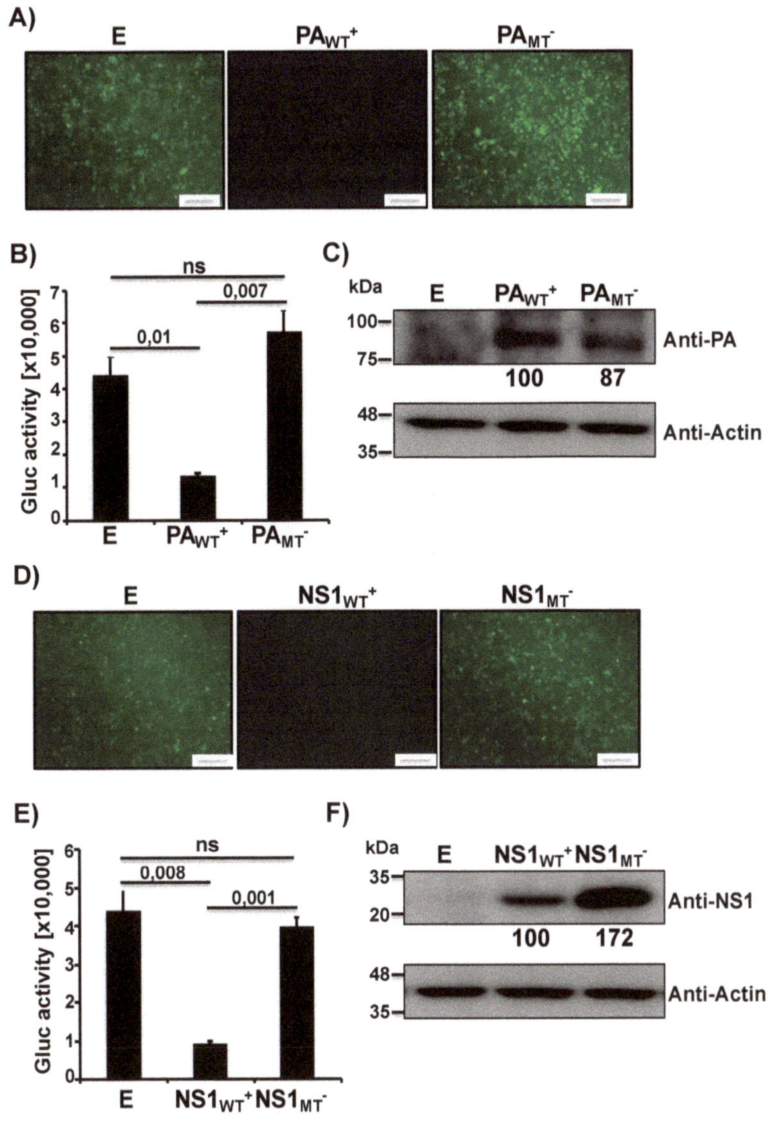

Figure 1. Ability of master donor virus (MDV) A/AA/6/60 live attenuated influenza vaccine (LAIV) PA and NS1 proteins to block host gene expression: Human 293T cells (5×10^5, 12-well plates, triplicates) were transiently co-transfected, using LPF2000, with expression plasmids encoding GFP and Gluc under the control of a chicken beta actin promoter (pCAGGS GFP and pCAGGS Gluc, respectively) together with pDZ plasmids encoding wild-type (WT^+) or mutant (MT^-) PA or NS1 proteins; or empty (**E**) plasmid as control. At 24 h post-transfection (p.t.), cells were analyzed by GFP expression (**A**,**D**) under a fluorescent microscope and by Gluc activity (**B**,**E**) from tissue culture supernatants (TCS) using a luminometer. Representative images are shown. Scale bar = 100 μm. Results represent the means and SDs of triplicate values. Protein expression from cell lysates was evaluated by Western blot (**C**,**F**) using specific antibodies for PA (**C**) or NS1 (**F**), or actin as the loading control. Molecular markers are indicated on the left. Western blots were quantified by densitometry using the software ImageJ. Relative band intensities (as described in Materials and Methods) are indicated. ns, not statistical.

Since both PA-X and NS1 are expressed together during viral infection, we next evaluated the inhibitory effects of PA-X and NS1 from MDV A/AA/6/60 LAIV in combination as well as the inhibitory effect if one or both had mutations that knocked out their function (Figure 2). To that end we co-transfected human 293T cells with the pCAGGS plasmids expressing GFP and Gluc along with a combination of either WT or mutant PA and NS1 pDZ expression plasmids. After 24 h p.t., GFP (Figure 2A) and Gluc (Figure 2B) expression levels were analyzed as described in Figure 1. The combination of both WT PA and NS1 proteins was the most effective at inhibiting host gene expression, while having one functional protein was still effective in significantly lowering reporter gene expression. Removing the functions of both viral proteins led to full expression of the reporter genes. Taken together these results demonstrate that we can tune the amount of inhibition of host cell protein synthesis in the MDV A/AA/6/60 LAIV by altering the ability of PA-X and NS1 proteins to inhibit host gene expression.

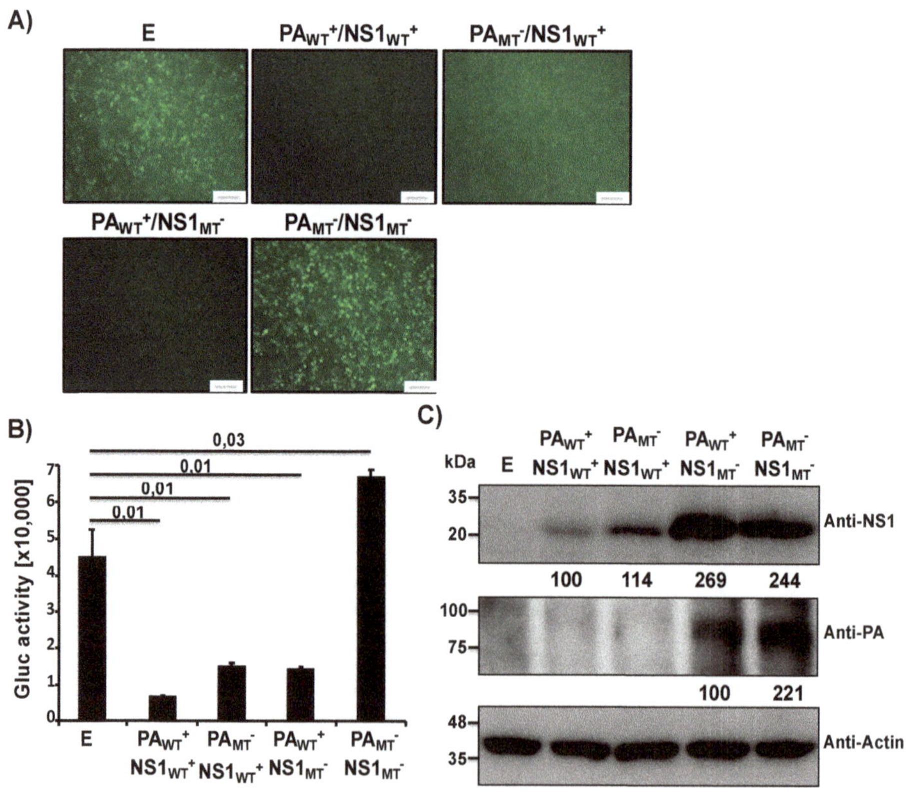

Figure 2. Ability of MDV A/AA/6/60 LAIV PA-X and NS1 proteins to block host gene expression in combination: Human 293T cells (5×10^5, 12-well plates, triplicates) were transiently co-transfected, using LPF2000, with pCAGGS GFP and pCAGGS Gluc plasmids together with the indicated combination of the pDZ plasmids encoding WT$^+$ or MT$^-$ PA and NS1 proteins; or empty (**E**) plasmid as control. At 24 h p.t., cells were analyzed for GFP expression under a fluorescent microscope (**A**) and for Gluc activity in the TCS using a luminometer (**B**). Representative images are shown. Scale bar = 100 µm. Results represent the means and SDs of triplicate values. Protein expression from cell lysates was evaluated by Western blot using specific antibodies for PA and NS1 proteins (**C**). Actin was used as the loading control. Molecular markers are noted on the left. Western blots were quantified by densitometry using the software ImageJ. Relative band intensities (as described in Materials and Methods) are indicated.

2.2. Generation of Recombinant WT and Mutant MDV A/AA/60 LAIVs Containing Modified PA and/or NS

In order to examine how inhibition of host gene expression affects the safety and immunogenicity of the MDV A/AA/6/60 LAIV, we built a plasmid-based reverse genetic system that we could manipulate to generate the recombinant viruses (Figure 3). To this end we started with the plasmids encoding the six internal proteins from the MDV A/AA/6/60 LAIV, which contain the 5 mutations (PB2 N265S; PB1 K391E, E581G, A661T; and NP D34G) that confer the *ca* and *ts* phenotype to the MDV A/AA/6/60 LAIV (Figure 3A) [10,11]. We then introduced synonymous mutations into the *PA* gene at the frameshift (Figure 3B), resulting in a PA expression plasmid that no longer generates PA-X (PA_{MT}^{-}) [15,18,19,21,23,39]. Last, we introduced two mutations into the *NS1* gene (F103S and M106I) to change amino acids that have been shown in other IAV strains to be necessary to bind CPSF30 and carry out the host gene expression inhibitory function of NS1 (Figure 3B) [35]. This set of plasmids allows us to look at the inhibitory effects of PA-X and NS1 separately and together during viral infection, allowing us to rescue the WT MDV A/AA/6/60 LAIV ($PA_{WT}^{+}/NS1_{WT}^{+}$), and three mutant MDV A/AA/6/60 LAIVs that have altered PA-X and/or NS1 functions ($PA_{WT}^{+}/NS1_{MT}^{-}$, $PA_{MT}^{-}/NS1_{WT}^{+}$, $PA_{MT}^{-}/NS1_{MT}^{-}$) (Figure 3A).

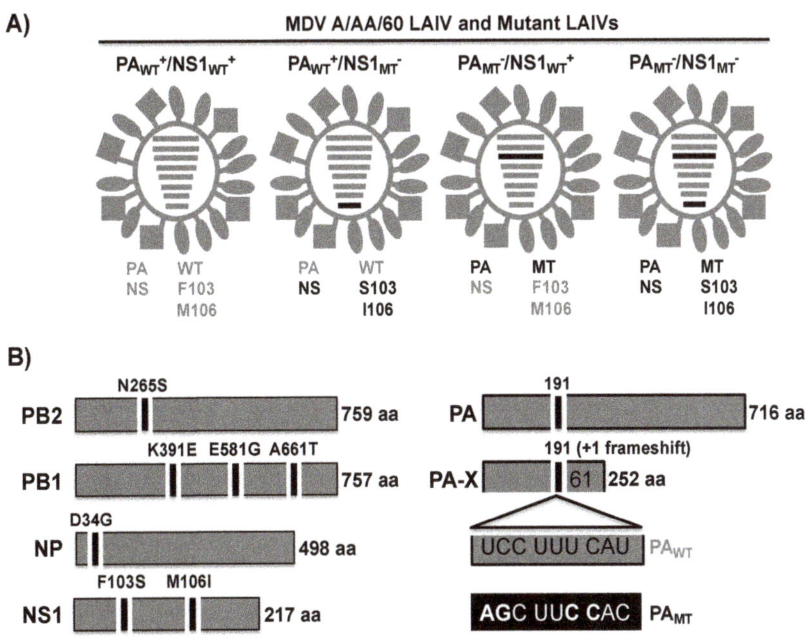

Figure 3. Schematic representation of the recombinant MDV A/AA/6/60 LAIVs. (**A**) Schematic representation of the MDV A/AA/6/60 WT and mutant LAIVs and their gene constellations: Recombinant A/AA/6/60 LAIVs with WT^{+} and MT^{-} segments are indicated with gray or black lines, respectively. $PA_{WT}^{+}/NS1_{WT}^{+}$: virus containing WT PA (UCC UUU CAU) and NS1 (F103/M106) proteins. $PA_{WT}^{+}/NS1_{MT}^{-}$: virus containing WT PA and mutant NS1 (S103/I106) proteins. $PA_{MT}^{-}/NS1_{WT}^{+}$: virus containing mutant PA (AGC UUC CAC) and WT NS1 proteins. $PA_{MT-}/NS1_{MT+}$: virus containing both PA and NS1 (mutant proteins. WT^{+} indicates viral proteins (PA and/or NS1) with the ability to inhibit host gene expression. MT^{-} indicates viral proteins (PA or NS1) unable to inhibit host gene expression. (**B**) Schematic representation of PB2, PB1, NP, PA, and NS1 viral proteins: A/AA/6/60 PB2, PB1, NP, and NS1 proteins (left) with the residues mutated to generate the MDV A/AA/6/60 LAIVs indicated. The WT and MT PA or PA-X viral proteins (right) and the mutations introduced into the frameshift motif (PA_{MT}) to abolish PA-X expression are shown. Numbers on the right indicate the amino acid (aa) length of the viral proteins.

2.3. Growth Kinetics of Recombinant WT and Mutant MDV A/AA/6/60 LAIVs

Next, we sought to determine how the recombinant WT MDV A/AA/6/60 LAIV and the mutant MDV A/AA/6/60 LAIVs with altered ability to inhibit host gene expression affected viral replication in vitro (Figure 4). To this end, we rescued five recombinant viruses: WT A/AA/6/60, MDV A/AA/6/60 LAIV ($PA_{WT}^{+}/NS1_{WT}^{+}$), and three mutant MDV A/AA/6/60 LAIVs ($PA_{WT}^{+}/NS1_{MT}^{-}$, $PA_{MT}^{-}/NS1_{WT}^{+}$, and $PA_{MT}^{-}/NS1_{MT}^{-}$), all of which express the HA and NA from pH1N1. In MDCK cells infected at low multiplicity of infection (MOI, 0.001), we observed that, compared to WT A/AA/6/60, the MDV A/AA/6/60 LAIV showed no growth at 39 °C and has reduced viral fitness at 37 °C, while maintaining robust growth at 33 °C, indicating that our reverse genetic system successfully rescued a functional *ts, ca* MDV A/AA/6/60 LAIV (Figure 4A). Next we examined the growth kinetics of the MDV A/AA/6/60 LAIV ($PA_{WT}^{+}/NS1_{WT}^{+}$), and the three mutant MDV A/AA/6/60 LAIVs ($PA_{WT}^{+}/NS1_{MT}^{-}$, $PA_{MT}^{-}/NS1_{WT}^{+}$, and $PA_{MT}^{-}/NS1_{MT}^{-}$) in MDCK and human A549 cells at 33 °C. In MDCK cells, all four viruses grow to similar titers but with different peaks of infection (Figure 4B). However, in A549 cells, the viruses lacking PA-X were significantly impaired in viral replication at later time points as compared to the two viruses encoding PA-X (Figure 4C). Moreover, when we examined the plaque morphology of the MDV A/AA/6/60 LAIV ($PA_{WT}^{+}/NS1_{WT}^{+}$) and the three mutant MDV A/AA/6/60 LAIVs ($PA_{WT}^{+}/NS1_{MT}^{-}$, $PA_{MT}^{-}/NS1_{WT}^{+}$, and $PA_{MT}^{-}/NS1_{MT}^{-}$), we saw that the viruses that do not express PA-X ($PA_{MT}^{-}/NS1_{WT}^{+}$ and $PA_{MT}^{-}/NS1_{MT}^{-}$) had plaques that were smaller compared to the two viruses encoding a WT PA-X (Figure 4C). In comparison, viruses which had a mutant NS1 and a functional PA-X ($PA_{WT}^{+}/NS1_{MT}^{-}$) had no change in plaque morphology as compared to the MDV A/AA6/60 LAIV ($PA_{WT}^{+}/NS1_{WT}^{+}$). These data suggest that PA-X is necessary for proper viral replication of the MDV A/AA/6/60 LAIV in vitro in both MDCK and A549 cells.

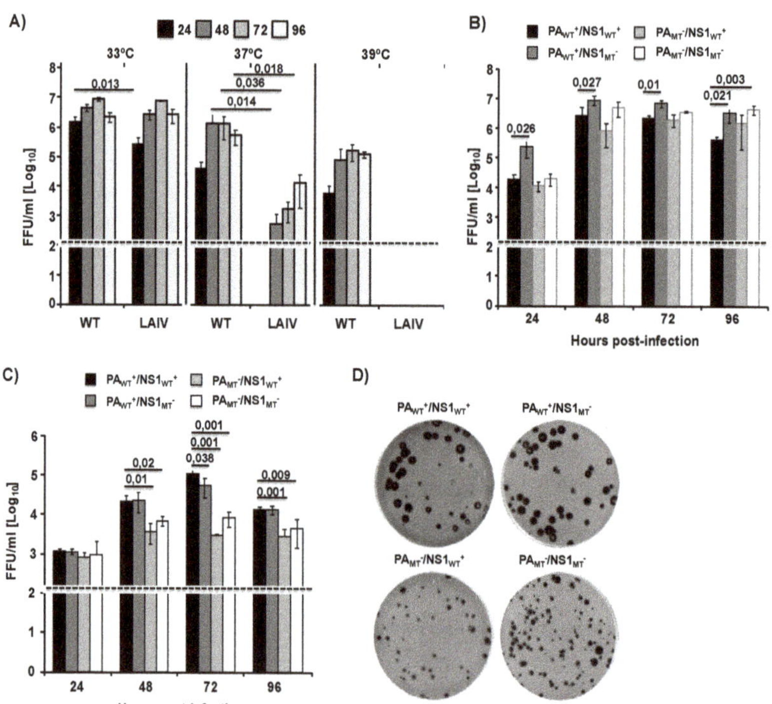

Figure 4. Multicycle growth kinetics and plaque assay of WT and mutant MDV A/AA/6/60 LAIVs. (**A**) Viral growth kinetics of WT A/AA/6/60 and MDV A/AA/6/60 LAIV at different temperatures: TCS from

MDCK cells (5 × 10^5, 12-well plates, triplicates) infected at low multiplicity of infection (MOI, 0.001) with WT A/AA/6/60 or MDV A/AA/6/60 LAIV at 33 °C, 37 °C, and 39 °C were analyzed at the indicated h p.i. (24, 48, 72, and 96) by immunofocus assay using an anti-NP mAb (HB-65). Data represent the means and SDs of the results determined from triplicate wells. The dashed line indicates the limit of detection (200 fluorescent forming units, FFU/mL). (**B**,**C**) Viral growth kinetics of WT MDV A/AA/6/60 LAIV (PA_{WT}^{+}/$NS1_{WT}^{+}$) and mutant MDV A/AA/6/60 LAIVs (PA_{WT}^{+}/$NS1_{MT}^{-}$, PA_{MT}^{-}/$NS1_{WT}^{+}$, and PA_{MT}^{-}/$NS1_{MT}^{-}$): MDCK (**B**) and A549 (**C**) cells (5 × 10^5, 12-well plates, triplicates) were infected (MOI of 0.001 and 0.025, respectively) with the indicated viruses and incubated at 33 °C. TCS were collected at the indicated h p.i. and viral titers were determined by immunofocus assay. Data represent the means and SDs of the results determined from triplicate wells. Dashed lines indicate the limit of detection (200 FFU/mL). *, $p < 0.05$ using Student's t test from Microsoft Excel. (**D**) Plaque assay: MDCK cells (1 × 10^6, 6-well plates) were infected with the indicated MDV A/AA/60 LAIVs and incubated at 33 °C for 3 days. Plaque phenotypes were visualized by immunostaining using the anti-NP mAb HB-65.

2.4. In Vivo Safety of MDV A/AA/6/60 LAIVs with Varying Ability to Inhibit Host Gene Expression

Next, we determined the safety profile of the mutant MDV A/AA/6/60 LAIVs (PA_{WT}^{+}/$NS1_{MT}^{-}$, PA_{MT}^{-}/$NS1_{WT}^{+}$, and PA_{MT}^{-}/$NS1_{MT}^{-}$) as compared to the WT MDV A/AA/6/60 LAIV (PA_{WT}^{+}/$NS1_{WT}^{+}$) in vivo (Figure 5). C57BL/6 mice were infected i.n. with 1 × 10^5 FFU of the indicated viruses and evaluated morbidity (body weight) (Figure 5A) and mortality (survival) (Figure 5B) for 14 days. We also examined viral titers in the lungs and nasal turbinates of C57BL/6-infected mice at 2 and 4 days p.i. (Figure 5C,D, respectively). C57BL/6 mice are a validated animal model of influenza infection, which has been used previously to evaluate IAV virulence, attenuation, and/or protection efficacy of new influenza vaccines [17,18,26,27,40,41]. None of the viruses caused weight loss, with all groups gaining weight over the course of the experiment (Figure 5A), although the weight variation was not directly compared to mock-infected animals. Moreover, as expected, all infected mice survived viral infection (Figure 5B). Notably, the WT MDV A/AA/6/60 LAIV, containing a mutated PA and a functional NS1 (PA_{MT}^{-}/$NS1_{WT}^{+}$) had undetectable viral titers in both the lungs (Figure 5C) and nasal turbinates (Figure 5D) at Day 4. These data, along with the previous results, suggest that removing PA-X results in a vaccine that is attenuated as compared to the WT MDV A/AA/6/60 LAIV, suggesting that a mutant MDV A/AA/6/60 LAIV lacking PA-X has a better safety profile than the currently MDV A/AA/6/60 LAIV.

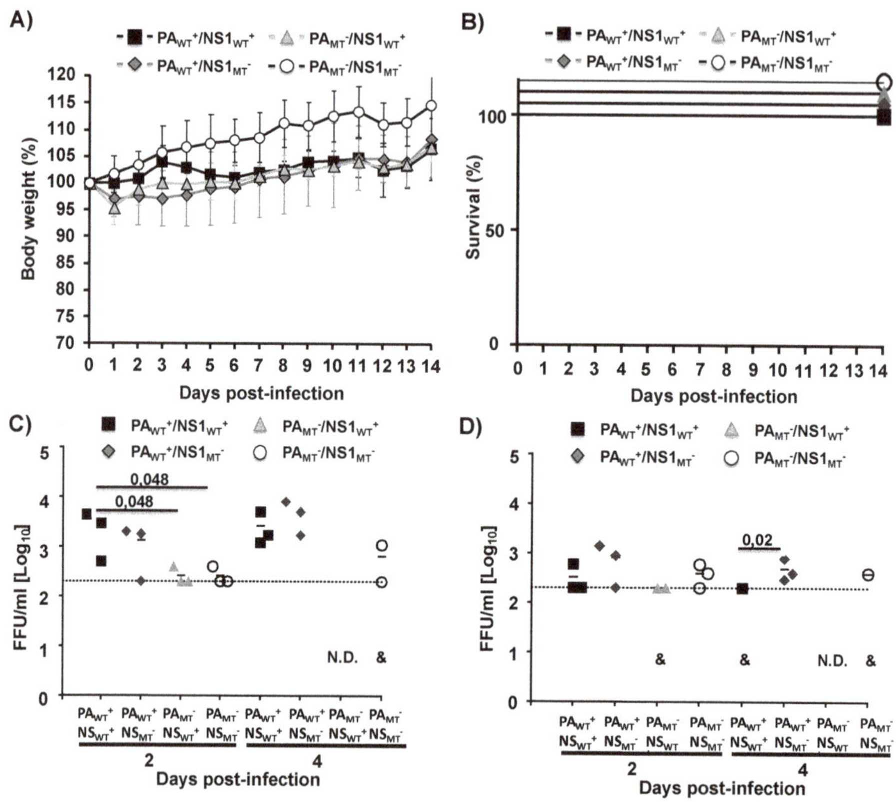

Figure 5. Safety profile of the WT MDV A/AA/6/60 LAIV and mutant MDV A/AA/6/60 LAIVs in vivo: Four-to-six-week-old female C57BL/6 mice ($n = 11$) were infected i.n. with 1×10^5 FFU of the WT MDV A/AA/6/60 LAIV ($PA_{WT}^+/NS1_{WT}^+$) or the mutant MDV A/AA/6/60 LAIVs ($PA_{WT}^+/NS1_{MT}^-$, $PA_{MT}^-/NS1_{WT}^+$, and $PA_{MT}^-/NS1_{MT}^-$). Body weight (**A**) and survival (**B**) were monitored for 14 days ($n = 5$). At Days 2 and 4 p.i., C57BL/6-infected mice were sacrificed ($n = 3$ per time point) and viral titers were determined in lungs (**C**) and nasal turbinates (**D**) by immunofocus assay using the anti-NP mAb HB-65. Symbols represent data from individual mice. Bars represent the mean for each group and indicates that the virus was only detected in one or two of the infected mice. N.D. indicates that virus was not detected. Dotted lines represent the limit of detection of the assay (200 FFU/mL).

2.5. Induction of Humoral Immune Responses by WT and Mutant MDV A/AA/6/60 LAIVs

Following the safety studies, we next determined if the mutant MDV A/AA/6/60 LAIVs ($PA_{WT}^+/NS1_{MT}^-$, $PA_{MT}^-/NS1_{WT}^+$, and $PA_{MT}^-/NS1_{MT}^-$) induced similar humoral immune responses than that of the WT MDV A/AA/6/60 LAIV ($PA_{WT}^+/NS1_{WT}^+$) (Figure 6). To this end, we collected blood from mice infected as shown in Figure 5, 21 days after infection and measured, using ELISA, the antibody responses against the viral H1 protein from pH1N1 (Figure 6A), as well as to all viral proteins using cell lysates from pH1N1-infected MDCK cells (Figure 6B). All the mutant MDV A/AA/6/60 LAIVs ($PA_{WT}^+/NS1_{MT}^-$, $PA_{MT}^-/NS1_{WT}^+$, and $PA_{MT}^-/NS1_{MT}^-$) induced an overall antibody response against the viral HA and other viral proteins that was comparable to that induced by the WT MDV A/AA/6/60 LAIV ($PA_{WT}^+/NS1_{WT}^+$) (Figure 6A,B, respectively). Next, we determined the ability of these antibodies to inhibit HA activity, using an hemagglutination inhibition (HAI) assay (Figure 6C). While all four MDV A/AA/6/60 LAIVs induced HAI antibodies, the MDV A/AA/6/60 LAIV containing both mutant PA and NS1 proteins ($PA_{MT}^{-}/NS1_{MT}^{-}$) induced significantly less HAI antibodies than the WT MDV A/AA/6/60 LAIV ($PA_{WT}^{+}/NS1_{WT}^{+}$).

A) B)

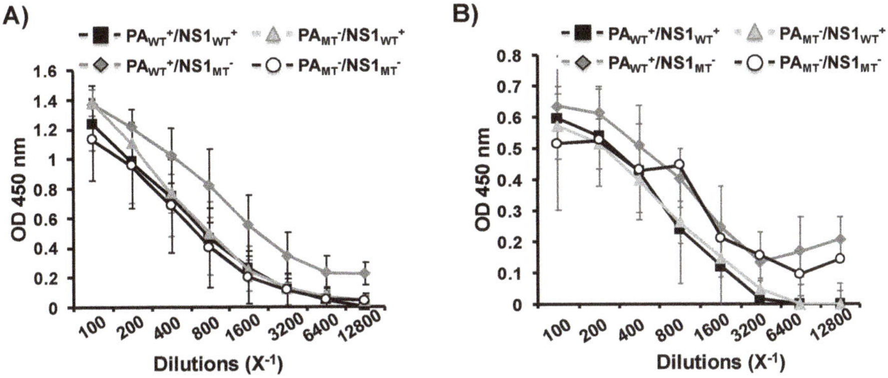

C)

Immunogenicity of LAIVs	
Virus	**HAI titer (SD)**
PA_{WT}^{+}/NS_{WT}^{+}	64 (22)
PA_{WT}^{+}/NS_{MT}^{-}	48 (18)
PA_{MT}^{-}/NS_{WT}^{+}	52 (26)
PA_{MT}^{-}/NS_{MT}^{-}	32 (11) *
PBS	≤ 10

Figure 6. Humoral responses induced by the MDV A/AA/6/60 LAIVs: Four-to-six-week-old female C57BL/6 mice (n = 5) were infected i.n. with 1×10^5 FFU with the WT MDV A/AA/6/60 LAIV ($PA_{WT}^{+}/NS1_{WT}^{+}$) or the mutant MDV A/AA/6/60 LAIVs ($PA_{WT}^{+}/NS1_{MT}^{-}$, $PA_{MT}^{-}/NS1_{WT}^{+}$, and $PA_{MT}^{-}/NS1_{MT}^{-}$) as indicated. At 21 days p.i., mice were bled, and sera were collected and evaluated for the presences of antibodies against recombinant pH1N1 HA (**A**) or total viral proteins using cell extracts from pH1N1 virus-infected MDCK cells (**B**) by ELISA. OD, optical density. (**C**) HAI titers were calculated from mouse serum. *, $p < 0.05$ ($PA_{WT}^{+}/NS1_{WT}^{+}$ vs. the other viruses) using a Student's t test from Microsoft Excel.

2.6. Protective Efficacy of MDV A/AA/6/60 LAIVs Against a Homologous pH1N1 Viral Challenge

Finally, we determined the protective efficacy of the three mutant MDV A/AA/6/60 LAIVs ($PA_{WT}^{+}/NS1_{MT}^{-}$, $PA_{MT}^{-}/NS1_{WT}^{+}$, and $PA_{MT}^{-}/NS1_{MT}^{-}$) as compared to the WT MDV A/AA/6/60 LAIV ($PA_{WT}^{+}/NS1_{WT}^{+}$) (Figure 7). To that end, mice were vaccinated with 1×10^5 FFU of the MDV A/AA/6/60 LAIVs and challenged 21 days p.i. with 1000X the mouse lethal dose 50 (MLD_{50}) of pH1N1 [42] and monitored for changes in body weight (Figure 7A) and survival (Figure 7B) for 14 days. We also evaluated the presence of the pH1N1 virus challenge in the lungs of infected animals at Days 2 and 4 post-challenge (Figure 7C). None of the vaccinated mice lost weight (Figure 7A) and all survived (Figure 7B) the lethal challenge with pH1N1, contrary to the situation of the mock (PBS)-vaccinated mice that all succumbed to viral infection. Mice that were vaccinated with the MDV A/AA/6/60 LAIV that had both mutant PA and NS1 ($PA_{MT}^{-}/NS1_{MT}^{-}$) showed significantly higher viral titers at Days 2 and 4 post-challenge compared to the mice vaccinated with the WT MDV A/AA/6/60 LAIV (Figure 7C), but significantly less than the PBS-treated control group, indicating that while there was a protective immune response (no mice died) it was not as robust, in agreement with the previous results showing reduced HAI antibodies (Figure 6C). The WT MDV A/AA/6/60 LAIV ($PA_{WT}^{+}/NS1_{WT}^{+}$) and the two mutant MDV A/AA/6/60 LAIVs containing only one mutant protein ($PA_{WT}^{+}/NS1_{MT}^{-}$ and

$PA_{MT}^{-}/NS1_{WT}^{+}$) showed a significant reduction in viral titers on both days as compared to the PBS control treated mice (Figure 7C). Taken together these results indicate that altering the ability of the WT MDV A/AA/6/60 LAIV to inhibit host protein synthesis does not abolish protective efficacy while improving its safety profile.

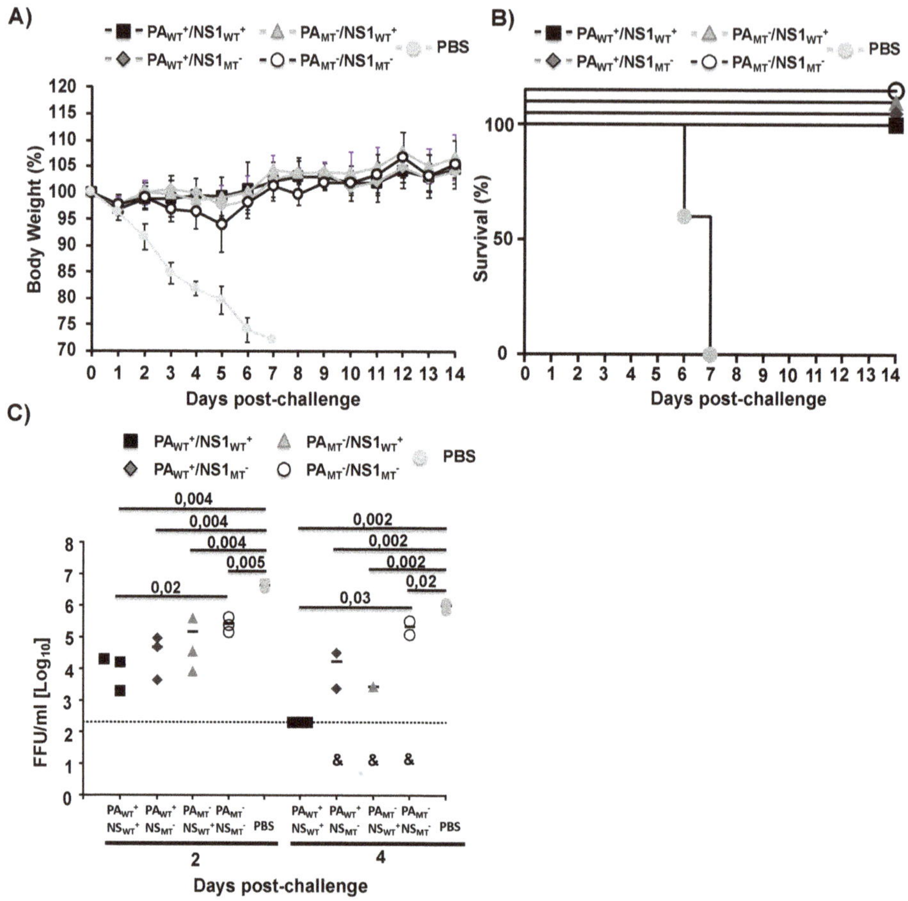

Figure 7. Protective efficacy of the MDV A/AA/6/60 LAIVs: Four-to-six-week-old female C57BL/6 mice ($n = 11$) were infected i.n. with 1×10^5 FFU of the WT MDV A/AA/6/60 LAIV ($PA_{WT}^{+}/NS1_{WT}^{+}$) or mutant MDV A/AA/6/60 LAIVs ($PA_{WT}^{+}/NS1_{MT}^{-}$, $PA_{MT}^{-}/NS1_{WT}^{+}$, $PA_{MT}^{-}/NS1_{MT}^{-}$), or PBS. At 21 days post-vaccination mice were challenged with 1000X MLD_{50} of pH1N1. Body weight (**A**) and mortality (**B**) were monitored for 14 days ($n = 5$). At Days 2 and 4 post-challenge, mice were sacrificed ($n = 3$ per time point) and viral titers in the lungs of infected animals were determined by immunofocus assay using the anti-NP mAb HB-65 (**C**). Symbols represent data from individual mice. Bars represent the geometric mean for each group and the indicated virus was detected only in one or two mice. The dotted line represents the limit of detection (200 FFU/mL).

3. Discussion

IAV infection requires a fine balance between inhibiting the host cell antiviral mechanisms by shutting down translation, and allowing enough of the host machinery active to allow for viral replication [16]. Since the discovery that PA-X and NS1 can both inhibit host gene expression, it has been seen that the different strains of IAV use each protein differently [16]. For example, pH1N1 uses PA-X exclusively for inhibiting host protein synthesis as its NS1 protein does not possess this function [17,18]. Furthermore, it has been shown that removing PA-X has different effects in different virus backbones. Removing PA-X expression in 1918 H1N1 [15], pH1N1 [39], and H9N2 [20] IAVs reduces viral pathogenicity, while removing PA-X in H5N1 IAV increases viral replication and pathogenicity [21]. These differing effects, together with our recent findings with pH1N1 [17,18] led us to believe that altering PA-X and/or NS1 function in the context of the WT MDV A/AA/6/60 LAIV would be a viable option to tune the safety, immunogenicity, and protection efficacy conferred by the LAIV. However, the possible side effects linked to changes in the ability of a new MDV A/AA/6/60 LAIV to modulate host innate immune responses, such as low immunogenicity, should be evaluated in other animal models of influenza infection. For instance, the implication in virus transmission should be tested in ferrets or guinea pigs, both well-stablished animal models of influenza virus infection and transmission [43–46].

The current MDV A/AA/6/60 LAIV has mutations mainly located in areas of the polymerase complex. There is always a concern that, with enough pressure, the MDV A/AA/6/60 LAIV could introduce mutations to compensate and revert back to a pathogenic strain [47]. By introducing mutations into additional viral genes (*PA* and *NS1*), the probability of reversion to a more virulent strain is decreased even further, adding to the increased safety profile desired for an MDV LAIV. However, the stability of the generated viruses should be tested using validated systems for IAV vaccine production. Additionally, by having another attenuation marker outside of the current PB2, PB1 and NP viral segments, there is less concern about the MDV A/AA/6/60 LAIV reassorting with a circulating strain.

In this work we demonstrate, for the first time, that the human MDV A/AA/6/60 LAIV has both active PA-X and NS1 proteins that can inhibit host gene expression (Figure 1). Building on our previous work, we also showed that the effects of these two viral proteins is additive in a reporter assay (Figure 2). We also demonstrate the feasibility of generating recombinant MDV A/AA/6/60 LAIVs with altered PA-X and/or NS1 proteins (Figure 3). Removing PA-X, however, showed to be detrimental for viral replication and resulted in a smaller plaque phenotype in cultured cells (Figure 4). Conversely, removing the inhibitory function of NS1 did not adversely affect viral replication of the MDV A/AA/6/60 LAIV (Figure 4). These trends were also seen in vivo. While all the MDV A/AA/6/60 LAIVs were still fully attenuated and did not result in any weight loss in mice, viruses lacking PA-X showed reduced viral replication in lungs and nasal turbinate, while those lacking a functional NS1 resembled the parental WT MDV A/AA/6/60 LAIV (Figure 5). Even with the reduced replication seen in some of these viruses, all of them were able to induce similar protective antibody responses (Figure 6) that provided full protection against a lethal homologous challenge with pH1N1 (Figure 7). However, while all of the MDV A/AA/6/60 LAIVs were protective against weight loss and mortality, there were differences in the amount of pH1N1 detected in the lungs of challenged mice at Days 2 and 4 p.i., with the parental WT MDV A/AA/6/60 LAIV showing the least amount of virus in the lungs and the double mutant MDV A/AA/6/60 LAIV (PA_{MT}^{-}/$NS1_{MT}^{-}$) showing the most pH1N1 viral replication (Figure 7).

Taken together our results show that removal of PA-X by modifying the frameshift sequence results in a vaccine that is fully protective but is further attenuated, suggesting that in a clinical setting this vaccine would be safer than the current MDV A/AA/6/60 LAIV in use. Thus, this new MDV A/AA/6/60 LAIV with reduced PA-X levels could be used for parts of the population where current LAIVs are not recommended, such as children under the age of two, immunocompromised individuals, the elderly, and pregnant adults. Furthermore, one of the major drawbacks for the current LAIV is that it can have adverse effects in children younger than two years of age, a cohort that would benefit greatly from an LAIV over the traditional chemically IIV. In addition, this safer MDV A/AA/6/60 LAIV lacking a functional PA-X will have another attenuation marker outside the current PB2, PB1, and NP segments, leading to less concerns of reversion to a WT virulent phenotype.

This and our previous work on the interplay of NS1 and PA-X in pH1N1 [17,18] show that modifying the functions of the PA-X and NS1 proteins is a viable way to fine tune the attenuation of vaccine candidates. Vaccination is the best way to prevent the spread of influenza and reduce the global health burden. However, in recent years the vaccine efficacy has been reported to be less than 50% leading to a large effort to find alternative vaccines that work better. This work demonstrates one method of increasing the safety profile of the current IAV LAIV without losing any of its protective efficacy that could be implemented with new universal vaccines strategies that are being currently developed for the treatment of influenza viral infections.

4. Materials and Methods

4.1. Cells and Viruses

Human embryonic kidney 293T (ATCC CRL-11268), human lung epithelial carcinoma A549 (ATCC CCL-185), and Madin-Darby canine kidney, MDCK (ATCC CCL-34) cells were grown and maintained in Dulbecco's modified Eagle's medium (DMEM) (Mediatech, Inc.) supplemented with 5% fetal bovine serum (FBS) (Atlanta Biologicals) and 1% penicillin (100 U/mL)–streptomycin (100 μg/mL)–2 mM L-glutamine, P-S-G (Mediatech, Inc.) at 37 °C in air enriched with 5% CO_2.

The recombinant wild-type (WT) A/California/4_NYICE_E3/2009 pH1N1 virus has been previously described [43]. The recombinant WT A/Ann Arbor/6/60 H2N2 (A/AA/6/60), MDV A/AA/6/60 LAIV, and mutant MDV A/AA/6/60 LAIVs were generated using plasmid-based reverse genetics as described below.

4.2. Plasmids

To generate the recombinant WT A/AA/6/60, MDV A/AA/6/60 LAIV, and mutant MDV A/AA/6/60 LAIVs, recombinant PB2 LAIV, PB1 LAIV, PA, and NP viral segments were synthesized de novo (Geneart) into a pMK plasmid. The recombinant M and NS segments were synthetized using Strings DNA Fragments technology (Geneart) and cloned into a pGEM-T plasmid (Promega). All the viral genes were synthetized with appropriate restriction sites at the 5′ and 3′ ends for subcloning into the ambisense plasmid pDZ [48]. To generate PB2 WT and PB1 WT, the *ts* mutations in PB2 (S265N), PB1 (E391K, G581E, and T661A) and NP (G34D) were introduced [49–51]. PA-X-deficient plasmids were created by mutating the frameshift motif from UCC UUU CAU (PA_{WT+}) to A<u>G</u>C UU<u>C</u> CA<u>C</u> (PA_{MT}^-) [18]. Two amino acid substitutions (F103S and M106I) [30,35] were introduced into the viral NS segment to mutate the NS1 protein. All the mutations were introduced by site-directed mutagenesis in the pMK or pGEM-T plasmids before subcloning into the ambisense pDZ plasmid. All plasmids were confirmed by sequencing (ACGT Inc.). Primers for the generation of the different plasmid constructs are available upon request.

4.3. Viral Rescues

Virus rescues were performed as we described previously [2,40,52]. Briefly, co-cultures (1:1) of 293T and MDCK cells in 6-well plates were co-transfected with 1 μg of each of the ambisense plasmids (pDZ-PB2, –PB1, $–PA_{WT}^+$ or PA_{MT}^-, –HA, –NP, –NA, –M, and NS_{WT}^+ or $NS1_{MT}^-$) using Lipofectamine 2000 (LPF2000, Invitrogen). At 12 h post-transfection (p.t.), the transfection medium was replaced with DMEM containing 0.3% bovine serum albumin (BSA), 1% P-S-G, and 0.5 μg/mL of N-tosyl-l-phenylalanine chloromethyl ketone (TPCK)-treated trypsin (Sigma). At 3–4 days p.t., tissue culture supernatants (TCS) were collected, clarified, and used to infect fresh MDCK cells. At 3 to 4 days post-infection (p.i.), recombinant viruses were plaque purified and scaled up in MDCK cells. Stocks were titrated by immunofocus assay (Fluorescence Focus Units (FFU)/mL)) on MDCK cells [2,52]. Virus stocks were confirmed by sequencing the PA and NS1 ORFs using purified total RNA (TRIzol reagent; Invitrogen) from infected MDCK cells according to the manufacturer's specifications.

4.4. Inhibition of Host Gene Expression

To evaluate the effect of MDV A/AA/6/60 LAIV NS1 and PA-X proteins on host protein synthesis, human 293T cells (2.5×10^5 cells/well, 24-well plate format, triplicates) were transiently co-transfected in suspension, using LPF2000, with 1 µg/well (500 ng/plasmid) of the indicated pDZ plasmids encoding WT or mutant PA (PA_{WT}^+ or PA_{MT}^-) proteins, WT or mutant NS1 ($NS1_{WT}^+$ or $NS1_{MT}^-$) proteins, together with 25 ng/well of pCAGGS plasmids expressing the green fluorescent protein (GFP) and Gaussia luciferase (Gluc). pDZ empty (E) plasmid was used as control or to complete 1 µg/well when only one NS or PA plasmid was added. At 24 h p.t., cells were analyzed for GFP expression under a fluorescence microscope and Gluc activity was quantified from TCS using a Biolux Gaussia luciferase reagent (New England BioLabs) and a Lumicount luminometer (Packard). The mean value with standard deviations (SDs) were calculated using Microsoft Excel software.

4.5. SDS-PAGE and Western Blot Analysis

Transfected cells were lysed in RIPA buffer (25 mM Tris HCl pH 7.6, 150 mM NaCl, 1% NP-40, 1% sodium deoxycholate, 0.1% SDS) and proteins were separated by denaturing electrophoresis using 10% SDS-polyacrylamide gels and transferred to a nitrocellulose membrane. Membranes were blocked for 1 h with 5% dried skim milk in phosphate-buffered saline (PBS) containing 0.1% Tween 20 (T-PBS) and incubated overnight at 4 °C with primary rabbit polyclonal antibodies (pAb) against NS1 [53] or a mix of mouse monoclonal antibodies (mAb) against PA (1F6, NR-19225; 5C5, NR-19226; and 8E10, NR-19224) obtained from BEI Resources. A mouse mAb against actin (A1978; Sigma) was used as an internal loading control. Horseradish peroxidase (HRP) secondary antibodies (GE Healthcare) against either mouse or rabbit immunoglobulins (Ig) were used to detect bound primary antibodies. Protein expression was detected with a SuperSignal West Femto maximum-sensitivity chemiluminescent substrate kit (Thermo Scientific) in accordance with the manufacturer's instructions. Protein bands were normalized to the level of beta-actin expression, and then the level of expression of the WT protein was considered 100%. In the case of Figure 2C, for the anti-PA analysis, $PA_{WT}+/NS1_{MT}-$ was considered 100%.

4.6. Viral Growth Kinetics

To assess virus growth kinetics in vitro, confluent monolayers of MDCK or A549 cells (4×10^5 cells/well, 12-well plate format, triplicates) were infected at a multiplicity of infection (MOI) of 0.001 (MDCK) or 0.025 (A549). After 1 h of virus adsorption at room temperature, cells were overlaid with DMEM containing 0.3% BSA, 1% P-S-G, and TPCK-treated trypsin (1 µg/mL for MDCK cells and 0.25 µg/mL for A549 cells) and incubated at 33 °C, 37 °C, or 39 °C for WT vs. LAIV and at 33 °C for the PA and NS1 mutants. At the indicated times post-infection, p.i. (24, 48, 72, and 96 h), TCS were collected and viral titers were determined by immunofocus assay (FFU/mL) as previously described [18,40,52]. Briefly, confluent wells of MDCK cells (10^4 cells/well, 96-well plate format, triplicates) were infected with 10-fold serial dilutions of TCS. At 12 h p.i., cells were fixed and permeabilized (4% formaldehyde and 0.5% Triton X-100 in PBS) for 15 min at room temperature. The cells were then incubated in blocking solution (2.5% BSA in PBS) for 1 h at room temperature and incubated with the influenza virus NP mAb HB-65 (ATCC) [18,40,52] for 1 h at 37 °C. After washing with PBS, cells were incubated with a fluorescein isothiocyanate (FITC)-conjugated rabbit anti-mouse IgG secondary antibody (Dako) for 1 h at 37 °C. NP-expressing positive cells were enumerated to determine the virus titer (FFU/mL). The mean value and SDs were calculated using Microsoft Excel software.

4.7. Plaque Assay and Immunostaining

Confluent monolayers of MDCK cells (10^6 cells/well, 6-well plate format) were infected for 1 h at room temperature, and after virus adsorption, cells were overlaid with agar and incubated at 33 °C. At 3 days p.i., cells were fixed with 4% paraformaldehyde for 15 min at room temperature.

After the overlays were removed, cells were permeabilized (0.5% Triton X-100 in PBS) for 15 min at room temperature and prepared for immunostaining as previously described [18,51], using the NP mAb HB-65 and vector kits (Vectastain ABC kit and DAB HRP substrate kit: Vector) following the manufacturer's specifications.

4.8. Mouse Experiments

Female 6-to-8-week-old C57BL/6 mice were purchased from the National Cancer Institute (NCI) and maintained in the animal care facility at the University of Rochester under specific-pathogen-free conditions. All animal protocols were approved by the University of Rochester Committee of Animal Resources and complied with the recommendations in the Guide for the Care and Use of Laboratory Animals of the National Research Council. Mice (n = 11/group) were anesthetized intraperitoneally (i.p.) with 2,2,2-tribromoethanol (Avertin; 240 mg/kg of body weight) and then infected intranasally (i.n.) with 30 µL of the indicated LAIVs. For challenge experiments mice (n = 11/group) were anesthetized i.p. and infected i.n. with 1000X the Median Lethal Dose (MLD_{50}) of pH1N1. Mice (n = 5/group) were monitored daily for morbidity (body weight loss) and mortality (survival) as previously described [17,26,27,36,40,41]. Mice showing 25% loss of their initial body weight were considered to have reached the experimental endpoint and were humanely euthanized. Virus replication was evaluated by determining viral titers in the lungs or nasal turbinates at 2 and 4 days p.i. To that end, three mice in each group were sacrificed, and lungs or nasal turbinates were extracted and homogenized. Virus titers were determined by immunofocus assay (FFU/mL) as indicated above.

4.9. Enzyme-Linked Immunosorbent Assay (ELISA)

Mouse sera were collected by submandibular bleeding at 21 days p.i. and evaluated for the presence of influenza virus-specific antibodies by ELISA [18,40]. Briefly, 96-well plates were coated for 16 h at 4 °C with lysates from mock- or pH1N1 WT virus-infected MDCK cells. Alternatively, plates were coated with pH1N1 HA (200 ng/well) (FR-180, International Reagent Resource) recombinant protein. After washing with PBS, coated wells were blocked with PBS containing 1% BSA and then incubated with 1:2 dilutions (starting dilution of 1:100) of mouse serum at 37 °C. After 1 h incubation, wells were washed with PBS and incubated with HRP-conjugated goat anti-mouse IgG (GE Healthcare) for 30 min at 37 °C. The reactions were developed with tetramethylbenzidine (TMB) substrate (BioLegend) for 10 min at room temperature, quenched with 2 N H_2SO_4, and read at 450 nm (Vmax kinetic microplate reader; Molecular Devices).

4.10. Hemagglutination inhibition (HAI) Assay

HAI assays were used to assess the presence of HA inhibition antibodies [18]. To that end, mouse sera were treated with receptor-destroying enzyme (RDE) (Denka Seiken) and heat inactivated for 30 min at 56 °C. Sera were then serially 2-fold diluted (starting dilution of 1:20) in 96-well V-bottom plates and mixed 1:1 with 4 hemagglutinating units (HAU) of pH1N1 WT for 60 min at room temperature. The HAI titers were determined by adding 0.5% turkey red blood cells (RBCs) to the virus-antibody mixtures for 30 min on ice.

Author Contributions: The project was conceptualized by A.N. and L.M.-S.; the experiments were performed by T.A.H., A.N., J.O. and K.C.; the manuscript was written by T.A.H. and A.N.; and the manuscript was edited by T.A.H., A.N., K.C. and L.M.-S. All authors have read and agreed to the published version of the manuscript.

Funding: This research was funded by the Department of Defense (DoD) Peer Reviewed Medical Research Program (PRMRP) grant W81XWH-18-1-0460 to LMS and by the T32 HL66988 to TAH.

Acknowledgments: The following reagents were obtained through BEI Resources, NIAID, NIH: Monoclonal anti-IAV PA clones 8E10 (NR-19224), 1F6 (NR-19225) and 5C5 (NR-19226). The following reagent was obtained through the International Reagent Resource (IRR): recombinant H1 HA from A/California/04/2009 (H1N1) pdm09 (FR-180). This research was partially funded by the New York Influenza Center of Excellence (NYICE), a member of the National Institute of Allergy and Infectious Diseases (NIAID), National Institutes of Health (NIH), Department of Health and Human Services, Centers of Excellence for Influenza Research and Surveillance (CEIRS) contract No. HHSN272201400005C (NYICE).

Conflicts of Interest: The authors declare no conflict of interest.

References

1. Shaw, M.L.; Palese, P. *Othomyxovirdae: the viruses and their replication*; Lippincott Williams and Wilkins: Philadelphia, PA, USA, 2007.
2. Baker, S.F.; Nogales, A.; Finch, C.; Tuffy, K.M.; Domm, W.; Perez, D.R.; Topham, D.J.; Martinez-Sobrido, L. Influenza A and B Virus Intertypic Reassortment through Compatible Viral Packaging Signals. *J. Virol.* **2014**, *88*, 10778–10791. [CrossRef] [PubMed]
3. Iuliano, A.D.; Roguski, K.M.; Chang, H.H.; Muscatello, D.J.; Palekar, R.; Tempia, S.; Cohen, C.; Gran, J.M.; Schanzer, D.; Cowling, B.J.; et al. Estimates of global seasonal influenza-associated respiratory mortality: A modelling study. *Lancet* **2018**, *391*, 1285–1300. [CrossRef]
4. Flannery, B.; Chung, J.R.; Belongia, E.A.; McLean, H.Q.; Gaglani, M.; Murthy, K.; Zimmerman, R.K.; Nowalk, M.P.; Jackson, M.L.; Jackson, L.A.; et al. Interim Estimates of 2017-18 Seasonal Influenza Vaccine Effectiveness—United States, February 2018. *MMWR Morb. Mortal. Wkly. Rep.* **2018**, *67*, 180–185. [CrossRef] [PubMed]
5. Mameli, C.; D'Auria, E.; Erba, P.; Nannini, P.; Zuccotti, G.V. Influenza vaccine response: Future perspectives. *Expert Opin. Biol. Ther.* **2018**, *18*, 1–5. [CrossRef]
6. Belongia, E.A.; Simpson, M.D.; King, J.P.; Sundaram, M.E.; Kelley, N.S.; Osterholm, M.T.; McLean, H.Q. Variable influenza vaccine effectiveness by subtype: A systematic review and meta-analysis of test-negative design studies. *Lancet Infect. Dis.* **2016**, *16*, 942–951. [CrossRef]
7. Darvishian, M.; Bijlsma, M.J.; Hak, E.; van den Heuvel, E.R. Effectiveness of seasonal influenza vaccine in community-dwelling elderly people: a meta-analysis of test-negative design case-control studies. *Lancet Infect. Dis.* **2014**, *14*, 1228–1239. [CrossRef]
8. Bergen, R.; Black, S.; Shinefield, H.; Lewis, E.; Ray, P.; Hansen, J.; Walker, R.; Hessel, C.; Cordova, J.; Mendelman, P.M. Safety of cold-adapted live attenuated influenza vaccine in a large cohort of children and adolescents. *Pediatr Infect. Dis. J.* **2004**, *23*, 138–144. [CrossRef]
9. Jin, H.; Chen, Z. Production of live attenuated influenza vaccines against seasonal and potential pandemic influenza viruses. *Curr. Opin. Virol.* **2014**, *6*, 34–39. [CrossRef]
10. Cox, N.J.; Kitame, F.; Kendal, A.P.; Maassab, H.F.; Naeve, C. Identification of sequence changes in the cold-adapted, live attenuated influenza vaccine strain, A/Ann Arbor/6/60 (H2N2). *Virology* **1988**, *167*, 554–567. [CrossRef]
11. Chan, W.; Zhou, H.; Kemble, G.; Jin, H. The cold adapted and temperature sensitive influenza A/Ann Arbor/6/60 virus, the master donor virus for live attenuated influenza vaccines, has multiple defects in replication at the restrictive temperature. *Virology* **2008**, *380*, 304–311. [CrossRef]
12. Snyder, M.H.; Betts, R.F.; DeBorde, D.; Tierney, E.L.; Clements, M.L.; Herrington, D.; Sears, S.D.; Dolin, R.; Maassab, H.F.; Murphy, B.R. Four viral genes independently contribute to attenuation of live influenza A/Ann Arbor/6/60 (H2N2) cold-adapted reassortant virus vaccines. *J. Virol.* **1988**, *62*, 488–495. [CrossRef] [PubMed]
13. Garcia-Sastre, A.; Egorov, A.; Matassov, D.; Brandt, S.; Levy, D.E.; Durbin, J.E.; Palese, P.; Muster, T. Influenza A virus lacking the NS1 gene replicates in interferon-deficient systems. *Virology* **1998**, *252*, 324–330. [CrossRef] [PubMed]
14. Hale, B.G.; Randall, R.E.; Ortin, J.; Jackson, D. The multifunctional NS1 protein of influenza A viruses. *J. Gen. Virol.* **2008**, *89*, 2359–2376. [CrossRef] [PubMed]
15. Jagger, B.W.; Wise, H.M.; Kash, J.C.; Walters, K.A.; Wills, N.M.; Xiao, Y.L.; Dunfee, R.L.; Schwartzman, L.M.; Ozinsky, A.; Bell, G.L.; et al. An overlapping protein-coding region in influenza A virus segment 3 modulates the host response. *Science* **2012**, *337*, 199–204. [CrossRef]
16. Nogales, A.; Martinez-Sobrido, L.; Topham, D.J.; DeDiego, M.L. Modulation of Innate Immune Responses by the Influenza A NS1 and PA-X Proteins. *Viruses* **2018**, *10*, 708. [CrossRef]
17. Nogales, A.; Martinez-Sobrido, L.; Chiem, K.; Topham, D.J.; DeDiego, M.L. Functional Evolution of the 2009 Pandemic H1N1 Influenza Virus NS1 and PA in Humans. *J. Virol.* **2018**, *92*. [CrossRef]

18. Nogales, A.; Rodriguez, L.; DeDiego, M.L.; Topham, D.J.; Martinez-Sobrido, L. Interplay of PA-X and NS1 Proteins in Replication and Pathogenesis of a Temperature-Sensitive 2009 Pandemic H1N1 Influenza A Virus. *J. Virol.* **2017**, *91*. [CrossRef]
19. Gao, H.; Sun, H.; Hu, J.; Qi, L.; Wang, J.; Xiong, X.; Wang, Y.; He, Q.; Lin, Y.; Kong, W.; et al. Twenty amino acids at the C-terminus of PA-X are associated with increased influenza A virus replication and pathogenicity. *J. Gen. Virol.* **2015**, *96*, 2036–2049. [CrossRef]
20. Gao, H.; Xu, G.; Sun, Y.; Qi, L.; Wang, J.; Kong, W.; Sun, H.; Pu, J.; Chang, K.C.; Liu, J. PA-X is a virulence factor in avian H9N2 influenza virus. *J. Gen. Virol.* **2015**, *96*, 2587–2594. [CrossRef]
21. Hu, J.; Mo, Y.; Wang, X.; Gu, M.; Hu, Z.; Zhong, L.; Wu, Q.; Hao, X.; Hu, S.; Liu, W.; et al. PA-X decreases the pathogenicity of highly pathogenic H5N1 influenza A virus in avian species by inhibiting virus replication and host response. *J. Virol.* **2015**, *89*, 4126–4142. [CrossRef]
22. Khaperskyy, D.A.; McCormick, C. Timing Is Everything: Coordinated Control of Host Shutoff by Influenza A Virus NS1 and PA-X Proteins. *J. Virol.* **2015**, *89*, 6528–6531. [CrossRef] [PubMed]
23. Khaperskyy, D.A.; Schmaling, S.; Larkins-Ford, J.; McCormick, C.; Gaglia, M.M. Selective Degradation of Host RNA Polymerase II Transcripts by Influenza A Virus PA-X Host Shutoff Protein. *PLoS Pathog.* **2016**, *12*, e1005427. [CrossRef] [PubMed]
24. Bavagnoli, L.; Cucuzza, S.; Campanini, G.; Rovida, F.; Paolucci, S.; Baldanti, F.; Maga, G. The novel influenza A virus protein PA-X and its naturally deleted variant show different enzymatic properties in comparison to the viral endonuclease PA. *Nucleic Acids Res.* **2015**, *43*, 9405–9417. [CrossRef] [PubMed]
25. Crotta, S.; Davidson, S.; Mahlakoiv, T.; Desmet, C.J.; Buckwalter, M.R.; Albert, M.L.; Staeheli, P.; Wack, A. Type I and type III interferons drive redundant amplification loops to induce a transcriptional signature in influenza-infected airway epithelia. *PLoS Pathog* **2013**, *9*, e1003773. [CrossRef]
26. DeDiego, M.L.; Nogales, A.; Lambert-Emo, K.; Martinez-Sobrido, L.; Topham, D.J. NS1 Protein Mutation I64T Affects Interferon Responses and Virulence of Circulating H3N2 Human Influenza A Viruses. *J. Virol.* **2016**, *90*, 9693–9711. [CrossRef]
27. Nogales, A.; Martinez-Sobrido, L.; Topham, D.J.; DeDiego, M.L. NS1 Protein Amino Acid Changes D189N and V194I Affect Interferon Responses, Thermosensitivity, and Virulence of Circulating H3N2 Human Influenza A Viruses. *J. Virol.* **2017**, *91*. [CrossRef]
28. Wolff, T.; Ludwig, S. Influenza viruses control the vertebrate type I interferon system: Factors, mechanisms, and consequences. *J. Interferon Cytokine Res.* **2009**, *29*, 549–557. [CrossRef]
29. Das, K.; Ma, L.C.; Xiao, R.; Radvansky, B.; Aramini, J.; Zhao, L.; Marklund, J.; Kuo, R.L.; Twu, K.Y.; Arnold, E.; et al. Structural basis for suppression of a host antiviral response by influenza A virus. *Proc. Natl. Acad. Sci. USA* **2008**, *105*, 13093–13098. [CrossRef]
30. Kochs, G.; Garcia-Sastre, A.; Martinez-Sobrido, L. Multiple anti-interferon actions of the influenza A virus NS1 protein. *J. Virol.* **2007**, *81*, 7011–7021. [CrossRef]
31. Noah, D.L.; Twu, K.Y.; Krug, R.M. Cellular antiviral responses against influenza A virus are countered at the posttranscriptional level by the viral NS1A protein via its binding to a cellular protein required for the 3′ end processing of cellular pre-mRNAS. *Virology* **2003**, *307*, 386–395. [CrossRef]
32. Ayllon, J.; Domingues, P.; Rajsbaum, R.; Miorin, L.; Schmolke, M.; Hale, B.G.; Garcia-Sastre, A. A single amino acid substitution in the novel H7N9 influenza A virus NS1 protein increases CPSF30 binding and virulence. *J. Virol.* **2014**, *88*, 12146–12151. [CrossRef] [PubMed]
33. Nemeroff, M.E.; Barabino, S.M.; Li, Y.; Keller, W.; Krug, R.M. Influenza virus NS1 protein interacts with the cellular 30 kDa subunit of CPSF and inhibits 3′end formation of cellular pre-mRNAs. *Mol. Cell* **1998**, *1*, 991–1000. [CrossRef]
34. Twu, K.Y.; Noah, D.L.; Rao, P.; Kuo, R.L.; Krug, R.M. The CPSF30 binding site on the NS1A protein of influenza A virus is a potential antiviral target. *J. Virol.* **2006**, *80*, 3957–3965. [CrossRef] [PubMed]
35. Steidle, S.; Martinez-Sobrido, L.; Mordstein, M.; Lienenklaus, S.; Garcia-Sastre, A.; Staheli, P.; Kochs, G. Glycine 184 in nonstructural protein NS1 determines the virulence of influenza A virus strain PR8 without affecting the host interferon response. *J. Virol.* **2010**, *84*, 12761–12770. [CrossRef] [PubMed]
36. Clark, A.M.; Nogales, A.; Martinez-Sobrido, L.; Topham, D.J.; DeDiego, M.L. Functional Evolution of Influenza Virus NS1 Protein in Currently Circulating Human 2009 Pandemic H1N1 Viruses. *J. Virol.* **2017**, *91*. [CrossRef]

37. Nogales, A.; Chauche, C.; DeDiego, M.L.; Topham, D.J.; Parrish, C.R.; Murcia, P.R.; Martinez-Sobrido, L. The K186E Amino Acid Substitution in the Canine Influenza Virus H3N8 NS1 Protein Restores Its Ability To Inhibit Host Gene Expression. *J. Virol.* **2017**, *91*. [CrossRef]
38. Hale, B.G.; Steel, J.; Medina, R.A.; Manicassamy, B.; Ye, J.; Hickman, D.; Hai, R.; Schmolke, M.; Lowen, A.C.; Perez, D.R.; et al. Inefficient control of host gene expression by the 2009 pandemic H1N1 influenza A virus NS1 protein. *J. Virol.* **2010**, *84*, 6909–6922. [CrossRef]
39. Lee, J.; Yu, H.; Li, Y.; Ma, J.; Lang, Y.; Duff, M.; Henningson, J.; Liu, Q.; Li, Y.; Nagy, A.; et al. Impacts of different expressions of PA-X protein on 2009 pandemic H1N1 virus replication, pathogenicity and host immune responses. *Virology* **2017**, *504*, 25–35. [CrossRef]
40. Nogales, A.; Baker, S.F.; Ortiz-Riano, E.; Dewhurst, S.; Topham, D.J.; Martinez-Sobrido, L. Influenza A Virus Attenuation by Codon Deoptimization of the NS Gene for Vaccine Development. *J. Virol.* **2014**, *88*, 10525–10540. [CrossRef]
41. Rodriguez, L.; Nogales, A.; Martinez-Sobrido, L. Influenza A Virus Studies in a Mouse Model of Infection. *J. Vis. Exp.* **2017**. [CrossRef]
42. Guo, H.; Santiago, F.; Lambert, K.; Takimoto, T.; Topham, D.J. T cell-mediated protection against lethal 2009 pandemic H1N1 influenza virus infection in a mouse model. *J. Virol.* **2011**, *85*, 448–455. [CrossRef] [PubMed]
43. Baker, S.F.; Guo, H.; Albrecht, R.A.; Garcia-Sastre, A.; Topham, D.J.; Martinez-Sobrido, L. Protection against lethal influenza with a viral mimic. *J. Virol.* **2013**, *87*, 8591–8605. [CrossRef] [PubMed]
44. Belser, J.A.; Pulit-Penaloza, J.A.; Maines, T.R. Ferreting Out Influenza Virus Pathogenicity and Transmissibility: Past and Future Risk Assessments in the Ferret Model. *Cold Spring Harb Perspect Med.* **2019**. [CrossRef] [PubMed]
45. Lowen, A.C.; Bouvier, N.M.; Steel, J. Transmission in the guinea pig model. *Curr. Top. Microbiol Immunol* **2014**, *385*, 157–183. [CrossRef]
46. Phipps, K.L.; Marshall, N.; Tao, H.; Danzy, S.; Onuoha, N.; Steel, J.; Lowen, A.C. Seasonal H3N2 and 2009 Pandemic H1N1 Influenza A Viruses Reassort Efficiently but Produce Attenuated Progeny. *J. Virol.* **2017**, *91*. [CrossRef]
47. Zhou, B.; Meliopoulos, V.A.; Wang, W.; Lin, X.; Stucker, K.M.; Halpin, R.A.; Stockwell, T.B.; Schultz-Cherry, S.; Wentworth, D.E. Reversion of Cold-Adapted Live Attenuated Influenza Vaccine into a Pathogenic Virus. *J. Virol.* **2016**, *90*, 8454–8463. [CrossRef]
48. Quinlivan, M.; Zamarin, D.; Garcia-Sastre, A.; Cullinane, A.; Chambers, T.; Palese, P. Attenuation of equine influenza viruses through truncations of the NS1 protein. *J. Virol.* **2005**, *79*, 8431–8439. [CrossRef]
49. Nogales, A.; Rodriguez, L.; Chauche, C.; Huang, K.; Reilly, E.C.; Topham, D.J.; Murcia, P.R.; Parrish, C.R.; Martinez-Sobrido, L. Temperature-Sensitive Live-Attenuated Canine Influenza Virus H3N8 Vaccine. *J. Virol.* **2017**, *91*. [CrossRef]
50. Rodriguez, L.; Nogales, A.; Reilly, E.C.; Topham, D.J.; Murcia, P.R.; Parrish, C.R.; Martinez Sobrido, L. A live-attenuated influenza vaccine for H3N2 canine influenza virus. *Virology* **2017**, *504*, 96–106. [CrossRef]
51. Nogales, A.; Martinez-Sobrido, L. Reverse Genetics Approaches for the Development of Influenza Vaccines. *Int. J. Mol. Sci.* **2016**, *18*, 20. [CrossRef]
52. Nogales, A.; Baker, S.F.; Martinez-Sobrido, L. Replication-competent influenza A viruses expressing a red fluorescent protein. *Virology* **2015**, *476*, 206–216. [CrossRef] [PubMed]
53. Solorzano, A.; Webby, R.J.; Lager, K.M.; Janke, B.H.; Garcia-Sastre, A.; Richt, J.A. Mutations in the NS1 protein of swine influenza virus impair anti-interferon activity and confer attenuation in pigs. *J. Virol.* **2005**, *79*, 7535–7543. [CrossRef] [PubMed]

pathogens

MDPI

Review

Formation and Maintenance of Tissue Resident Memory CD8+ T Cells after Viral Infection

David J. Topham *, Emma C. Reilly, Kris Lambert Emo and Mike Sportiello

David H. Smith Center for Vaccine Biology and Immunology, Department of Microbiology and Immunology, University of Rochester Medical Center, Rochester, NY 14642, USA; emma_reilly@urmc.rochester.edu (E.C.R.); kris_lambert@urmc.rochester.edu (K.L.E.); michael_sportiello@urmc.rochester.edu (M.S.)

* Correspondence: david_tophamr@urmc.rochester.edu; Tel.: +1-585-273-1400

Received: 25 September 2019; Accepted: 16 October 2019; Published: 18 October 2019

Abstract: Tissue resident memory (T_{RM}) CD8 T cells comprise a memory population that forms in peripheral, non-lymphoid tissues after an infection that does not recirculate into the bloodstream or other tissues. T_{RM} cells often recognize conserved peptide epitopes shared among different strains of a pathogen and so offer a protective role upon secondary encounter with the same or related pathogens. Several recent studies have begun to shed light on the intrinsic and extrinsic factors regulating T_{RM}. In addition, work is being done to understand how canonical "markers" of T_{RM} actually affect the function of these cells. Many of these markers regulate the generation or persistence of these T_{RM} cells, an important point of study due to the differences in persistence of T_{RM} between tissues, which may impact future vaccine development to cater towards these important differences. In this review, we will discuss recent advances in T_{RM} biology that may lead to strategies designed to promote this important protective immune subset.

Keywords: T cell; memory; tissue resident; immune response

1. Introduction

As early as 1994, Walter Gerhard's lab demonstrated that protection from a serologically distinct strain of influenza in immune mice was largely due to cross-reactive CD4 and CD8 T cells, with CD8 T cells having the greatest effect in the airways and lung [1]. This form of immunity, called heterosubtypic protection, was effective but short-lived and waned over several months. In 2001, David Woodland's lab used a different respiratory pathogen, Sendai (parainfluenza) virus, to demonstrate that functional virus-specific CD8 T cells persisted in the airways and lungs. However, they observed a similar decline in numbers over a year after infection [2], and the number of virus-specific CD8 T cells in the airways correlated with protection. These studies pre-date the identification of cell surface markers used to define T_{RM} cells, although Hogan et al. did measure CD69 expression [2]. In the same year, David Masopust and Leo Lefrancois used major histocompatibility complex (MHC) tetramers carrying viral or bacterial-specific peptides to show that after a systemic infection, pathogen-specific memory CD8 T cells localized to all peripheral non-lymphoid after infection [3]. In 2004, inspired by this data, our lab used an influenza virus model to demonstrate that Very Late Antigen-1 (VLA-1) (the CD49a/CD29 $\alpha1\beta1$ integrin heterodimer) was essential to maintain virus-specific CD8 memory T cells in the airways and lung tissues [4]. In that same study, we found virus-specific CD8 T cells expressing VLA-1 in every peripheral tissue examined, showing that infection of a tissue is not a prerequisite for T_{RM} localization. Overall, these results indicate that infection results in populations of CD49a+, CD69+, CD8+ memory T cells that reside in peripheral tissues: tissue resident memory T cells. [4]. Given the importance of this memory T cell population to secondary immune protection, here we review recent advances in the understanding of T_{RM} and call for further research that will lead to strategies designed to improve local tissue immunity.

2. Markers Used to Define T_{RM}

Expression of several cell surface proteins have been found to be common among T_{RM} from different tissues. Absence of chemokine receptor 7 (CCR7) and expression of CD69 were identified early [5,6], and CD49a or VLA-1 came next [4]. At the time, reagents to study mouse CD49a were limited in availability, so this marker was not initially widely adopted. In 2005, CD103 expression was found to define a population of memory T cells in the tonsil [7] and in 2010, the Bevan lab associated CD103 with T_{RM} in the brain [8]. Since then, CD69 and CD103 expression have been the most widely used markers of T_{RM}, although more recent studies in both mice and humans have demonstrated the importance of CD49a. It may be necessary to use these three markers (CD69, CD49a, CD103) in combination when studying T_{RM} from different tissues, as it is unlikely to be a homogenous population as previously assumed. Many studies of function and gene expression of T_{RM} have relied on either CD69 or CD103 alone or in combination [9,10]. Although CD69 is critical for T_{RM} in some peripheral tissues (e.g., kidney), it does not appear to be a universal requirement for T_{RM} in all sites [11]. Our lab has recently discovered that using CD49a and CD103 in combination identifies up to four potentially distinct memory T cell subsets in the lungs and airways (including trachea), although cells expressing only CD69 and CD103 do not appear to persist long-term [12]. Using all three markers may be necessary to paint the full picture of cell diversity, function, and gene expression profiles (Table 1).

Table 1. Cell Surface Expression of Markers on Memory CD8 T cell subsets in the airways and lung tissue.

Memory type	Abbreviation	CD3	TCRb	CD8	CD44	CD62L	CD69	CD49a	CD103
Effector	EM	+	+	+	high	low	−	−	−
Epithelial T_{RM}	eT_{RM}	+	+	+	high	low	+	+	+
Interstitial T_{RM}	iT_{RM}	+	+	+	high	low	+	+	−
Epithelial Effector	eeT_{RM}	+	+	+	high	low	+	−	+

Note: "+" indicates positive cell surface expression; "−" indicates the cell population is negative for cell surface expression "high" and "low" refer to the levels of cell surface expression.

3. Functions of Memory T Cell Markers

The cell surface proteins used to define T_{RM} and other memory subsets are not just markers; they have functions, though relatively little has been done to discern these functions [13]. Naïve T cells lack expression of CD49a and CD69 and have only low levels of CD103 on their surface. Instead, they can be defined by high expression of CD62L (L-selectin), low expression of CD44, and expression of CCR7 [14]. CD62L is a ligand for a receptor expressed on high endothelial venules to help direct naïve T cells to secondary lymphoid organs (SLOs) [15]. CCR7 serves a similar purpose by receiving CCL19 and CCL21 chemokine signals coming from SLOs [16,17]. CD44 binds hyaluronic acids in the extracellular matrix of almost all tissues and may itself be important for regulation of T cell motility, positioning, and retention in peripheral and lymphoid tissues [18,19].

"Central memory T cells" (T_{CM}) are $CD44^{high}$, $CD62L^{high}$, and retain CCR7 expression [14] and lymphoid homing potential. "Effector memory T cells" (T_{EM}) are $CD44^{high}$, $CD62L^{low}$, and $CCR7^{-}$, and preferentially circulate though the bloodstream and into through peripheral non-lymphoid tissues [14]. T_{EM} form one of the four memory CD8 T cell subsets found in lung and airways, and do not express CD49a, CD103, and CD69. CD49a/VLA-1 is the only known integrin ligand for collagen IV, uniquely located in the lamina densa of barrier tissues, though it has been shown to also bind collagen I [20]. It provides anti-apoptotic signals to the T cells when bound to collagen [21] and has proven essential for maintaining T_{RM} in peripheral tissues. T cells that express CD49a/VLA-1 tend to localize close to or within the epithelial cell layer where collagen IV is abundant [22]. CD69 is an S1P1 antagonist, limiting signals that direct lymphocytes into draining lymphatics [23,24]. It is also a ligand for Galectin-1 (Gal-1) and may have a role in cell–cell and cell–matrix interactions [25].

CD103 is a ligand for E-cadherin, which is most widely expressed in the junctions between epithelial cells at mucosal and barrier sites [26]. Both CD8 T cells and dendritic cells (DC) can also express this ligand for CD103, two more sources of potential cell–cell interaction [27]. CD103 is also expressed on a subset of DC in the airways that serve to sample the epithelium for antigen by interaction with cell junctions [28], possibly playing a similar role for CD8 T_{RM} cells in epithelial surveillance. Expression of these adhesion and migratory receptors may position memory T cell subsets in different locations in the mucosa depending on ligand availability (Figure 1, Table 2)

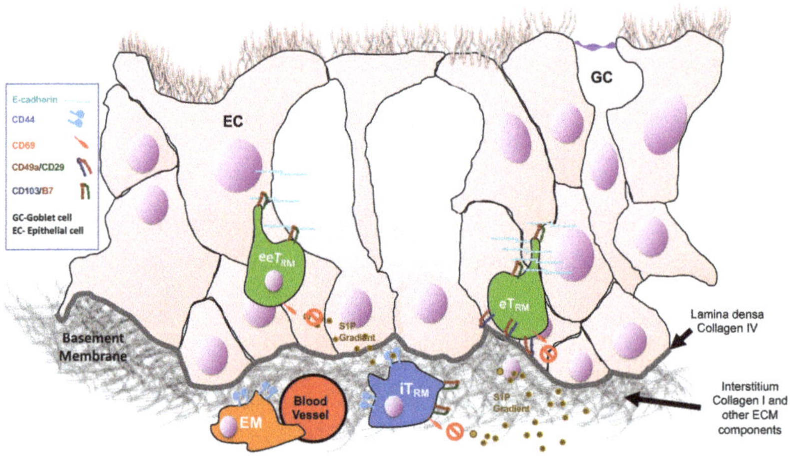

Figure 1. Proposed tissue localization of CD8+ memory T cell subsets within epithelial tissue. Based on the expression of adhesion molecules for ECM (CD49a, CD44) and, epithelial cells (CD103), and the location of their ligands, the memory cell subsets may be positioned in different tissue microenvironments.

Table 2. Select markers of memory CD8 T cell subsets and their functions.

Marker	Function	Presence on T_{RM}
CD44	Binds hyaluronic acids. May serve roles in motility and retention in both peripheral and lymphoid tissues.	+/High
CD62L	Ligand for high endothelial venules on SLO.	−/Low
CCR7	Chemokine receptor for S1P1.	−/Low
CD49a	Mediates adhesion to type I and IV collagen.	Positive on most T_{RM}
CD69	Early activation marker. S1P1R antagonist. Binds Gal-1.	Positive on most T_{RM}
CD103	Binds E-Cadherin.	Very often present, though can be tissue-dependent.

Note: "+" indicates positive cell surface expression; "−" indicated negative cell surface expression.

4. Formation of CD8 T_{RM} and Interactions with Other Cells in the Tissue

Virus-specific CD8 cells that express CD49a do not appear in the tissue until after the infection is cleared [4]. In influenza infection, CD8 T cells in the tissue express CD49a, CD69, and CD103 by day 14, suggesting cells with a T_{RM} phenotype develop relatively early as the tissue recovers from infection [12]. There has been much speculation regarding when and where T_{RM} cells form: Takamura demonstrated that in areas of repair after influenza infection, there are repair-associated memory depots (RAMD) containing populations of keratin-5+ cells that express alpha-V integrins [29]. Alpha-V integrins can activate latent TGF−β in the tissue, a necessary cytokine for T_{RM} to express CD49a and CD103 [30,31]. This suggests the T_{RM} cells form directly in the tissues they take residence in.

However, this hypothesis does not explain the early observations that pathogen-specific T_{RM} can be found in all peripheral tissues examined regardless of whether they were directly infected [4]. Experiments using FTY720 to block S1P signaling and inhibit lymph node (LN) egress resulted in an accumulation of CD8 T cells bearing the T_{RM} phenotype in the lymph nodes (Topham unpublished). In such an experiment it is difficult to discern in situ differentiation of T_{RM} in the LN from drainage of the cells out of the tissues, but it does question whether T_{RM} cells solely develop in the recovering tissue, and also suggests there may be a window after infection during which T_{RM} cells disperse systemically to peripheral tissues. Takamura's parabiosis experiments suggest that mice need to be paired by day 6 after infection to show distribution of virus specific cells to lungs and other tissues of both mice [29]. However, more focused tracking experiments will need to be done to concretely answer this question.

Additional signals in the tissue may be needed to generate T_{RM}. Using a push-pull strategy, the Kohlmeier lab recently demonstrated that encounter with antigen in the tissue may be required for T_{RM} localization [32]. Using an intramuscular infection model and intranasal administration of CpG or CpG plus nucleoprotein (NP) peptide, only the animals given local peptide developed substantial NP-specific T_{RM} cells in the respiratory tract [32]. Antigen encounter has been shown to down-regulate CCR7 on recently activated T cells perhaps explaining the need for antigen to be present for optimal T_{RM} formation [33–35].

A role for IL-10 has also been indicated in the production of monocyte-derived TGF−β; a critical mediator for upregulation of CD103 (and possibly CD49a) in several studies [36–38]. These experiments used various adjuvants and antigen in vivo, and an in vitro model of T cell activation to induce an immune response. With this system, T cells exposed to IL10 and TGF−β very early in priming had the highest levels of CD103, suggesting T_{RM} precursors could be generated in the LN [38]. This scenario of early exposure to these cytokines is less likely to occur in models of infection given the slow kinetics (days) and levels of IL-10, which are not high early in a flu infection [39], but antigen-bearing monocytes releasing Transforming Growth Factor (TGF)−β directly in relation to T_{RM} development has not been studied in the context of infection.

It is also possible that CD103 could be expressed prior to other T_{RM} markers, and inhibition of CD103 leads to reduced numbers of T_{RM}, consistent with a role in the accumulation in tissues [23]. Although these papers did not look at the function of CD49a, in a gut model of T_{RM}, TGF−β was shown to be important for expression of CD49 on a4b7+ gut T_{RM} [37] In an influenza infection model, another stimulatory signal for T_{RM} formation comes from 4–1BB [40]. 4–1BB is a TNF family receptor and endogenous signals from antigen-bearing 4–1BB ligand expressing cells in the tissue enhance T_{RM} formation while exogenous administration of 4–1BB ligand further expands the number of T_{RM} formed [40].

5. Metabolic Changes Associated with T_{RM}

Tissues like skin, gut, and respiratory mucosa are very different environments than Secondary Lymphoid Organ (SLO)s and blood. The availability of nutrients such as glucose and oxygen vary compared to SLOs. The metabolic pathways utilized by T cells has been shown to vary with their state of activation [41]. As with most eukaryotic cells, naïve T cells primarily derive energy from oxidative phosphorylation in the mitochondria [42] to make adenosine triphosphate (ATP). T cell activation drives T cell metabolism toward aerobic glycolysis, the pentose phosphate pathway, and glutaminolysis [43]. These changes are associated with reprogramming of the metabolic transcriptome [44,45]. The different circulating memory T cells subsets utilize distinct metabolic programs. For example, central memory T cells primarily use oxidative phosphorylation, much like naïve T cells [42,46]. Effector memory T cells that can circulate through peripheral tissues use a more balanced combination of oxidative phosphorylation and glycolysis and have greater mitochondrial mass that makes them more comparable to effector T cells [47].

Peripheral tissues, especially barrier tissues and microcompartments contained within these offer different glucose levels, pH, and structural components, all of which can affect T cell metabolism and survival. T_{RM} cells are very active in the "resting" state as they constantly perform surveillance of the cells surrounding them for evidence of pathogen invasion. T_{RM} cells intimately interact with the components of their environment including epithelial cells and the extracellular matrix. T_{RM} cells abundantly express lipid receptors on their cell surface to provide them capacity for lipid uptake and metabolism [48]. These include fatty acid binding proteins FABP4, FABP5 to enhance lipolysis, low density lipid receptor, ApoE, and the CD36 scavenger receptor. These receptors facilitate uptake of free fatty acids (FFA) from the environment and these FFA can be used in oxidative phosphorylation and in fatty acid oxidation (FAO) [49]. T_{RM} can accumulate FFA in droplets within the cytoplasm just like T_{EM} are able to do [50], and these droplets can be coupled to distinct mitochondria upon activation [51,52]. T_{RM} may rely on traditional oxidative phosphorylation and glycolysis under "resting" conditions, and activate FAO by accessing the stored droplets upon encounter with pathogens and antigen recognition, however, this has yet to be directly tested.

6. Mechanisms that Suppress T_{RM} Activation May Have other Functions Important to T_{RM} Establishment and Maintenance

CD8 T_{RM} cells are armed and ready to protect against secondary infections. They can express high levels of Granzyme B giving them cytotoxic potential, secrete IFN−γ and TNF−α when activated, as well as a laundry list of chemokines to jumpstart immune cell recruitment [41]. Therefore, accidental or bystander activation of T_{RM} could result in serious, unwanted tissue damage. There are a number of inhibitory receptors expressed by CD8 T_{RM} that likely serve to keep these cells at bay until bona fide antigen recognition occurs through the TCR. Brain T_{RM} cells express PD1, a receptor associated with T cell exhaustion and inability to activate [53,54]. Blockade of PD-1 ligand in tumor settings greatly enhances T cell elimination of cancer cells [55]. How this suppression is overcome in T_{RM} during an infection is not known and may require a combination of "activating" signals. P2RX7 is a purinergic receptor expressed by T_{RM} [56]. Low ATP levels and signals through P2RX7 can promote T_{RM} survival by stimulating activation and metabolism. High signals, on the other hand, can drive T cell apoptosis [56], making the balance of signals in the environment of T_{RM} critical. T_{RM} cells also express CD244, a receptor previously described on Natural Killer (NK) cells [57]. Signals through CD244 can be either inhibitory or activating depending on intracellular levels of the adapter protein Signaling Lymphocytic Activation Molecule (SLAM) associated protein (SAP) [58]. SAP levels in T_{RM} are not known, but this knowledge could further indicate how these, and other receptors regulate T_{RM} maintenance and function. Collectively, these inhibitory molecules may keep T_{RM} in a state of readiness to respond, but inhibiting aberrant activation of T_{RM} by non-specific environmental cues.

7. Concluding Remarks

T_{RM} cells represent a very specialized subset of memory T cells that can be highly protective during secondary encounters with a previously seen pathogen. While not discussed here, they also seem to be important in controlling tumor cells and CD103 expression has been suggested as a predictor of tumor prognosis [59]. They have specialized functions that permit them to perform surveillance and protection at a number of barrier sites, including but not limited to skin, gut, respiratory tract, and female reproductive tract. However, many questions remain in our understanding of this unique cell subset. For example, we don't fully appreciate why they seem to wane with time in some tissues, while they are continuously renewed in others. We also do not have effective, clinically feasible strategies of generating, maintaining, or improving their numbers. Live attenuated vaccines do generate a population of T_{RM} cells, however at a weakened capacity compared with natural infection [60,61]. Additionally, to date, T_{RM} cells are considered as one uniform population of cells and we believe that research examining T_{RM} subsets is lacking. For example, do all T_{RM} cells express CD69, CD49a, or CD103? Do populations exist that express different combinations of these markers? What are the genetic and epigenetic programs that drive their phenotype(s)? What are the mechanisms of controlling motility, persistence, and micro-localization within different tissues? In what ways, if at all, are T_{RM} cells specific for viruses different than those specific for bacteria, fungi, cancer, or other pathologies? These are all important questions the field needs to address to construct a more holistic model through which we cannot only understand memory and immunity, but use them to develop more effective vaccines and therapies.

Author Contributions: D.J.T. organized the review and wrote the text; E.C.R. edited the manuscript and contributed data; K.L.E. edited the manuscript; M.S. edited the manuscript and drew Figure 1.

Funding: This research was funded by National Institutes of Health/National Institute of Allergy and Infectious Disease grant number 2P01-AI102851 and National Institute of General Medical Sciences Medical Scientist Training Program T32 GM007356.

Conflicts of Interest: The authors declare no conflict of interest.

References

1. Liang, S.; Mozdzanowska, K.; Palladino, G.; Gerhard, W. Heterosubtypic immunity to influenza type A virus in mice. Effector mechanisms and their longevity. *J. Immunol.* **1994**, *152*, 1653–1661. [PubMed]
2. Hogan, R.J.; Usherwood, E.J.; Zhong, W.; Roberts, A.A.; Dutton, R.W.; Harmsen, A.G.; Woodland, D.L. Activated antigen-specific CD8+ T cells persist in the lungs following recovery from respiratory virus infections. *J. Immunol.* **2001**, *166*, 1813–1822. [CrossRef] [PubMed]
3. Masopust, D.; Vezys, V.; Marzo, A.L.; Lefrancois, L. Preferential localization of effector memory cells in nonlymphoid tissue. *Science* **2001**, *291*, 2413–2417. [CrossRef] [PubMed]
4. Ray, S.J.; Franki, S.N.; Pierce, R.H.; Dimitrova, S.; Koteliansky, V.; Sprague, A.G.; Doherty, P.C.; de Fougerolles, A.R.; Topham, D.J. The collagen binding alpha1beta1 integrin VLA-1 regulates CD8 T cell-mediated immune protection against heterologous influenza infection. *Immunity* **2004**, *20*, 167–179. [CrossRef]
5. Teijaro, J.R.; Turner, D.; Pham, Q.; Wherry, E.J.; Lefrancois, L.; Farber, D.L. Cutting edge: Tissue-retentive lung memory CD4 T cells mediate optimal protection to respiratory virus infection. *J. Immunol.* **2011**, *187*, 5510–5514. [CrossRef]
6. Sallusto, F.; Lenig, D.; Forster, R.; Lipp, M.; Lanzavecchia, A. Two subsets of memory T lymphocytes with distinct homing potentials and effector functions [see comments]. *Nature* **1999**, *401*, 708–712. [CrossRef]
7. Woodberry, T.; Suscovich, T.J.; Henry, L.M.; August, M.; Waring, M.T.; Kaur, A.; Hess, C.; Kutok, J.L.; Aster, J.C.; Wang, F.; et al. Alpha E beta 7 (CD103) expression identifies a highly active, tonsil-resident effector-memory CTL population. *J. Immunol.* **2005**, *175*, 4355–4362. [CrossRef]
8. Wakim, L.M.; Woodward-Davis, A.; Bevan, M.J. Memory T cells persisting within the brain after local infection show functional adaptations to their tissue of residence. *Proc. Natl. Acad. Sci. USA* **2010**, *107*, 17872–17879. [CrossRef]
9. Kumar, B.V.; Ma, W.; Miron, M.; Granot, T.; Guyer, R.S.; Carpenter, D.J.; Senda, T.; Sun, X.; Ho, S.H.; Lerner, H.; et al. Human Tissue-Resident Memory T Cells Are Defined by Core Transcriptional and Functional Signatures in Lymphoid and Mucosal Sites. *Cell Rep.* **2017**, *20*, 2921–2934. [CrossRef]
10. Mackay, L.K.; Minnich, M.; Kragten, N.A.; Liao, Y.; Nota, B.; Seillet, C.; Zaid, A.; Man, K.; Preston, S.; Freestone, D.; et al. Hobit and Blimp1 instruct a universal transcriptional program of tissue residency in lymphocytes. *Science* **2016**, *352*, 459–463. [CrossRef]
11. Walsh, D.A.; Borges da Silva, H.; Beura, L.K.; Peng, C.; Hamilton, S.E.; Masopust, D.; Jameson, S.C. The Functional Requirement for CD69 in Establishment of Resident Memory CD8(+) T Cells Varies with Tissue Location. *J. Immunol.* **2019**, *203*, 946–955. [CrossRef] [PubMed]
12. Reilly, E.C.; Lambert-Emo, K.; Reilly, N.; Chaves, F.A.; Oakes, P.; Topham, D.J. TRM Integrins CD103 and CD49a Differentially Support Adherence and Motility After Resolution of Influenza Virus Infection. *Proc. Natl. Acad. Sci. USA* **2019**, in press.

13. Topham, D.J.; Reilly, E.C. Tissue-Resident Memory CD8(+) T Cells: From Phenotype to Function. *Front. Immunol.* **2018**, *9*, 515. [CrossRef] [PubMed]
14. Sallusto, F.; Geginat, J.; Lanzavecchia, A. Central memory and effector memory T cell subsets: function, generation, and maintenance. *Annu. Rev. Immunol.* **2004**, *22*, 745–763. [CrossRef] [PubMed]
15. Spertini, O.; Luscinskas, F.W.; Kansas, G.S.; Munro, J.M.; Griffin, J.D.; Gimbrone, M.A., Jr.; Tedder, T.F. Leukocyte adhesion molecule-1 (LAM-1, L-selectin) interacts with an inducible endothelial cell ligand to support leukocyte adhesion. *J. Immunol.* **1991**, *147*, 2565–2573.
16. Sallusto, F.; Kremmer, E.; Palermo, B.; Hoy, A.; Ponath, P.; Qin, S.; Forster, R.; Lipp, M.; Lanzavecchia, A. Switch in chemokine receptor expression upon TCR stimulation reveals novel homing potential for recently activated T cells. *Eur. J. Immunol.* **1999**, *29*, 2037–2045. [CrossRef]
17. Debes, G.F.; Hopken, U.E.; Hamann, A. In vivo differentiated cytokine-producing CD4(+) T cells express functional CCR7. *J. Immunol.* **2002**, *168*, 5441–5447. [CrossRef]
18. Lazaar, A.L.; Albelda, S.M.; Pilewski, J.M.; Brennan, B.; Pure, E.; Panettieri, R.A., Jr. T lymphocytes adhere to airway smooth muscle cells via integrins and CD44 and induce smooth muscle cell DNA synthesis. *J. Exp. Med.* **1994**, *180*, 807–816. [CrossRef]
19. Mrass, P.; Kinjyo, I.; Ng, L.G.; Reiner, S.L.; Pure, E.; Weninger, W. CD44 mediates successful interstitial navigation by killer T cells and enables efficient antitumor immunity. *Immunity* **2008**, *29*, 971–985. [CrossRef]
20. Bank, I.; Book, M.; Ware, R. Functional role of VLA-1 (CD49A) in adhesion, cation-dependent spreading, and activation of cultured human T lymphocytes. *Cell Immunol.* **1994**, *156*, 424–437. [CrossRef]
21. Richter, M.V.; Topham, D.J. The alpha1beta1 integrin and TNF receptor II protect airway CD8+ effector T cells from apoptosis during influenza infection. *J. Immunol.* **2007**, *179*, 5054–5063. [CrossRef] [PubMed]
22. Richter, M.; Ray, S.J.; Chapman, T.J.; Austin, S.J.; Rebhahn, J.; Mosmann, T.R.; Gardner, H.; Kotelianski, V.; de Fougerolles, A.R.; Topham, D.J. Collagen distribution and expression of collagen-binding alpha1beta1 (VLA-1) and alpha2beta1 (VLA-2) integrins on CD4 and CD8 T cells during influenza infection. *J. Immunol.* **2007**, *178*, 4506–4516. [CrossRef]
23. Lee, Y.T.; Suarez-Ramirez, J.E.; Redman, J.M.; Aguila, C.C.; Hadley, G.A.; Cauley, L.S. CD69 and CD103 cooperatively regulate CD8 T cell responses in the lungs after viral infection. *J. Immunol.* **2009**, *182*.
24. Lee, Y.T.; Suarez-Ramirez, J.E.; Wu, T.; Redman, J.M.; Bouchard, K.; Hadley, G.A.; Cauley, L.S. Environmental and antigen receptor-derived signals support sustained surveillance of the lungs by pathogen-specific cytotoxic T lymphocytes. *J. Virol.* **2011**, *85*, 4085–4094. [CrossRef] [PubMed]
25. Cibrian, D.; Sanchez-Madrid, F. CD69: from activation marker to metabolic gatekeeper. *Eur. J. Immunol.* **2017**, *47*, 946–953. [CrossRef] [PubMed]
26. Pauls, K.; Schon, M.; Kubitza, R.C.; Homey, B.; Wiesenborn, A.; Lehmann, P.; Ruzicka, T.; Parker, C.M.; Schon, M.P. Role of integrin alphaE(CD103)beta7 for tissue-specific epidermal localization of CD8+ T lymphocytes. *J. Invest. Dermatol.* **2001**, *117*, 569–575. [CrossRef] [PubMed]
27. Cepek, K.L.; Rimm, D.L.; Brenner, M.B. Expression of a candidate cadherin in T lymphocytes. *Proc. Natl. Acad. Sci. USA* **1996**, *93*, 6567–6571. [CrossRef]
28. Sung, S.S.J.; Fu, S.M.; Rose, C.E.; Gaskin, F.; Ju, S.T.; Beaty, S.R. A Major Lung CD103 (E)- 7 Integrin-Positive Epithelial Dendritic Cell Population Expressing Langerin and Tight Junction Proteins. *J. Immunol.* **2006**, *176*, 2161–2172. [CrossRef]
29. Takamura, S.; Yagi, H.; Hakata, Y.; Motozono, C.; McMaster, S.R.; Masumoto, T.; Fujisawa, M.; Chikaishi, T.; Komeda, J.; Itoh, J.; et al. Specific niches for lung-resident memory CD8+ T cells at the site of tissue regeneration enable CD69-independent maintenance. *J. Exp. Med.* **2016**, *213*, 3057–3073. [CrossRef]
30. Casey, K.A.; Fraser, K.A.; Schenkel, J.M.; Moran, A.; Abt, M.C.; Beura, L.K.; Lucas, P.J.; Artis, D.; Wherry, E.J.; Hogquist, K.; et al. Antigen-independent differentiation and maintenance of effector-like resident memory T cells in tissues. *J. Immunol.* **2012**, *188*, 4866–4875. [CrossRef]
31. Mackay, L.K.; Wynne-Jones, E.; Freestone, D.; Pellicci, D.G.; Mielke, L.A.; Newman, D.M.; Braun, A.; Masson, F.; Kallies, A.; Belz, G.T.; et al. T-box Transcription Factors Combine with the Cytokines TGF-beta and IL-15 to Control Tissue-Resident Memory T Cell Fate. *Immunity* **2015**, *43*, 1101–1111. [CrossRef] [PubMed]
32. McMaster, S.R.; Wein, A.N.; Dunbar, P.R.; Hayward, S.L.; Cartwright, E.K.; Denning, T.L.; Kohlmeier, J.E. Pulmonary antigen encounter regulates the establishment of tissue-resident CD8 memory T cells in the lung airways and parenchyma. *Mucosal Immunol.* **2018**, *11*, 1071–1078. [CrossRef] [PubMed]

33. Bromley, S.K.; Thomas, S.Y.; Luster, A.D. Chemokine receptor CCR7 guides T cell exit from peripheral tissues and entry into afferent lymphatics. *Nat. Immunol.* **2005**, *6*, 895–901. [CrossRef] [PubMed]
34. Debes, G.F.; Arnold, C.N.; Young, A.J.; Krautwald, S.; Lipp, M.; Hay, J.B.; Butcher, E.C. Chemokine receptor CCR7 required for T lymphocyte exit from peripheral tissues. *Nat. Immunol.* **2005**, *6*, 889–894. [CrossRef]
35. Jennrich, S.; Lee, M.H.; Lynn, R.C.; Dewberry, K.; Debes, G.F. Tissue exit: a novel control point in the accumulation of antigen-specific CD8 T cells in the influenza a virus-infected lung. *J. Virol.* **2012**, *86*, 3436–3445. [CrossRef]
36. Thompson, E.A.; Darrah, P.A.; Foulds, K.E.; Hoffer, E.; Caffrey-Carr, A.; Norenstedt, S.; Perbeck, L.; Seder, R.A.; Kedl, R.M.; Lore, K. Monocytes Acquire the Ability to Prime Tissue-Resident T Cells via IL-10-Mediated TGF-beta Release. *Cell. Rep.* **2019**, *28*, 1127–1135 e1124. [CrossRef]
37. Zhang, N.; Bevan, M.J. Transforming growth factor-beta signaling controls the formation and maintenance of gut-resident memory T cells by regulating migration and retention. *Immunity* **2013**, *39*, 687–696. [CrossRef]
38. Nath, A.P.; Braun, A.; Ritchie, S.C.; Carbone, F.R.; Mackay, L.K.; Gebhardt, T.; Inouye, M. Comparative analysis reveals a role for TGF-beta in shaping the residency-related transcriptional signature in tissue-resident memory CD8+ T cells. *PLoS ONE* **2019**, *14*, e0210495. [CrossRef]
39. McKinstry, K.K.; Strutt, T.M.; Buck, A.; Curtis, J.D.; Dibble, J.P.; Huston, G.; Tighe, M.; Hamada, H.; Sell, S.; Dutton, R.W.; et al. IL-10 deficiency unleashes an influenza-specific Th17 response and enhances survival against high-dose challenge. *J. Immunol.* **2009**, *182*, 7353–7363. [CrossRef]
40. Zhou, A.C.; Batista, N.V.; Watts, T.H. 4-1BB Regulates Effector CD8 T Cell Accumulation in the Lung Tissue through a TRAF1-, mTOR-, and Antigen-Dependent Mechanism to Enhance Tissue-Resident Memory T Cell Formation during Respiratory Influenza Infection. *J. Immunol.* **2019**, *202*, 2482–2492. [CrossRef]
41. Konjar, S.; Veldhoen, M. Dynamic Metabolic State of Tissue Resident CD8 T Cells. *Front. Immunol.* **2019**, *10*, 1683. [CrossRef] [PubMed]
42. Fox, C.J.; Hammerman, P.S.; Thompson, C.B. Fuel feeds function: energy metabolism and the T-cell response. *Nat. Rev. Immunol.* **2005**, *5*, 844–852. [CrossRef] [PubMed]
43. Doedens, A.L.; Phan, A.T.; Stradner, M.H.; Fujimoto, J.K.; Nguyen, J.V.; Yang, E.; Johnson, R.S.; Goldrath, A.W. Hypoxia-inducible factors enhance the effector responses of CD8(+) T cells to persistent antigen. *Nat. Immunol.* **2013**, *14*, 1173–1182. [CrossRef] [PubMed]
44. Macintyre, A.N.; Finlay, D.; Preston, G.; Sinclair, L.V.; Waugh, C.M.; Tamas, P.; Feijoo, C.; Okkenhaug, K.; Cantrell, D.A. Protein kinase B controls transcriptional programs that direct cytotoxic T cell fate but is dispensable for T cell metabolism. *Immunity* **2011**, *34*, 224–236. [CrossRef]
45. Wang, R.; Dillon, C.P.; Shi, L.Z.; Milasta, S.; Carter, R.; Finkelstein, D.; McCormick, L.L.; Fitzgerald, P.; Chi, H.; Munger, J.; et al. The transcription factor Myc controls metabolic reprogramming upon T lymphocyte activation. *Immunity* **2011**, *35*, 871–882. [CrossRef]
46. Maekawa, Y.; Ishifune, C.; Tsukumo, S.; Hozumi, K.; Yagita, H.; Yasutomo, K. Notch controls the survival of memory CD4+ T cells by regulating glucose uptake. *Nat. Med.* **2015**, *21*, 55–61. [CrossRef]
47. Van der Windt, G.J.; Everts, B.; Chang, C.H.; Curtis, J.D.; Freitas, T.C.; Amiel, E.; Pearce, E.J.; Pearce, E.L. Mitochondrial respiratory capacity is a critical regulator of CD8+ T cell memory development. *Immunity* **2012**, *36*, 68–78. [CrossRef]
48. Fahrer, A.M.; Konigshofer, Y.; Kerr, E.M.; Ghandour, G.; Mack, D.H.; Davis, M.M.; Chien, Y.H. Attributes of gammadelta intraepithelial lymphocytes as suggested by their transcriptional profile. *Proc. Natl. Acad. Sci. USA* **2001**, *98*, 10261–10266. [CrossRef]
49. Pan, Y.; Tian, T.; Park, C.O.; Lofftus, S.Y.; Mei, S.; Liu, X.; Luo, C.; O'Malley, J.T.; Gehad, A.; Teague, J.E.; et al. Survival of tissue-resident memory T cells requires exogenous lipid uptake and metabolism. *Nature* **2017**, *543*, 252–256. [CrossRef]
50. Konjar, S.; Frising, U.C.; Ferreira, C.; Hinterleitner, R.; Mayassi, T.; Zhang, Q.; Blankenhaus, B.; Haberman, N.; Loo, Y.; Guedes, J.; et al. Mitochondria maintain controlled activation state of epithelial-resident T lymphocytes. *Sci. Immunol.* **2018**, *3*. [CrossRef]
51. Benador, I.Y.; Veliova, M.; Liesa, M.; Shirihai, O.S. Mitochondria Bound to Lipid Droplets: Where Mitochondrial Dynamics Regulate Lipid Storage and Utilization. *Cell Metab* **2019**, *29*, 827–835. [CrossRef] [PubMed]
52. Benador, I.Y.; Veliova, M.; Mahdaviani, K.; Petcherski, A.; Wikstrom, J.D.; Assali, E.A.; Acin-Perez, R.; Shum, M.; Oliveira, M.F.; Cinti, S.; et al. Mitochondria Bound to Lipid Droplets Have Unique Bioenergetics,

Composition, and Dynamics that Support Lipid Droplet Expansion. *Cell Metab.* **2018**, *27*, 869–885 e866. [CrossRef] [PubMed]
53. Prasad, S.; Hu, S.; Sheng, W.S.; Chauhan, P.; Singh, A.; Lokensgard, J.R. The PD-1: PD-L1 pathway promotes development of brain-resident memory T cells following acute viral encephalitis. *J. Neuroinflammation* **2017**, *14*, 82. [CrossRef] [PubMed]
54. Shwetank; Frost, E.L.; Mockus, T.E.; Ren, H.M.; Toprak, M.; Lauver, M.D.; Netherby-Winslow, C.S.; Jin, G.; Cosby, J.M.; Evavold, B.D.; et al. PD-1 Dynamically Regulates Inflammation and Development of Brain-Resident Memory CD8 T Cells During Persistent Viral Encephalitis. *Front. Immunol.* **2019**, *10*, 783. [CrossRef]
55. Peggs, K.S.; Quezada, S.A. PD-1 blockade: promoting endogenous anti-tumor immunity. *Expert Rev. Anticancer Ther.* **2012**, *12*, 1279–1282. [CrossRef]
56. Borges da Silva, H.; Beura, L.K.; Wang, H.; Hanse, E.A.; Gore, R.; Scott, M.C.; Walsh, D.A.; Block, K.E.; Fonseca, R.; Yan, Y.; et al. The purinergic receptor P2RX7 directs metabolic fitness of long-lived memory CD8(+) T cells. *Nature* **2018**, *559*, 264–268. [CrossRef] [PubMed]
57. Konjar, S.; Ferreira, C.; Blankenhaus, B.; Veldhoen, M. Intestinal Barrier Interactions with Specialized CD8 T Cells. *Front. Immunol.* **2017**, *8*, 1281. [CrossRef] [PubMed]
58. Waggoner, S.N.; Kumar, V. Evolving role of 2B4/CD244 in T and NK cell responses during virus infection. *Front. Immunol.* **2012**, *3*, 377. [CrossRef] [PubMed]
59. Edwards, J.; Wilmott, J.S.; Madore, J.; Gide, T.N.; Quek, C.; Tasker, A.; Ferguson, A.; Chen, J.; Hewavisenti, R.; Hersey, P.; et al. CD103(+) Tumor-Resident CD8(+) T Cells Are Associated with Improved Survival in Immunotherapy-Naive Melanoma Patients and Expand Significantly During Anti-PD-1 Treatment. *Clin. Cancer Res. Offic. J. Am. Assoc. Cancer Res.* **2018**, *24*, 3036–3045. [CrossRef]
60. Horvath, K.M.; Brighton, L.E.; Herbst, M.; Noah, T.L.; Jaspers, I. Live attenuated influenza virus (LAIV) induces different mucosal T cell function in nonsmokers and smokers. *Clin. Immunol.* **2012**, *142*, 232–236. [CrossRef]
61. Xu, X.; Wang, H.; Liu, Y.; Wang, Y.; Zeng, L.; Wu, K.; Wang, J.; Ma, F.; Xu, W.; Yin, Y.; et al. Mucosal immunization with the live attenuated vaccine SPY1 induces humoral and Th2-Th17-regulatory T cell cellular immunity and protects against pneumococcal infection. *Infect. Immun.* **2015**, *83*, 90–100. [CrossRef] [PubMed]

pathogens

MDPI

Review

Role of Memory B Cells in Hemagglutinin-Specific Antibody Production Following Human Influenza A Virus Infection

Mark Y. Sangster *, Phuong Q. T. Nguyen and David J. Topham

David H. Smith Center for Vaccine Biology and Immunology, Department of Microbiology and Immunology, University of Rochester Medical Center, Rochester, NY 14642, USA; phuong_nguyen@urmc.rochester.edu (P.Q.T.N.); david_topham@urmc.rochester.edu (D.J.T.)

* Correspondence: mark_sangster@urmc.rochester.edu; Tel.: +1-585-273-3109

Received: 23 August 2019; Accepted: 26 September 2019; Published: 28 September 2019

Abstract: When influenza A virus infects an immune individual, preexisting memory B cell (MBC) activation and rapid anamnestic antibody production plays a key role in viral clearance. The most effective neutralizing antibodies target the antigenically variable head of the viral hemagglutinin (HA); antibodies against the conserved HA stalk provide broader but less potent protection. In this review, we provide a comprehensive picture of an adult's HA-specific antibody response to influenza virus infection. The process is followed from preexisting HA-specific MBC activation and rapid production of anti-HA antibodies, through to germinal center seeding and adaptation of the response to novel features of the HA. A major focus of the review is the role of competition between preexisting MBCs in determining the character of the HA-reactive antibody response. HA novelty modifies this competition and can shift the response from the immunodominant head to the stalk. We suggest that antibodies resulting from preexisting MBC activation are important regulators of anti-HA antibody production and play a role in positive selection of germinal center B cells reactive to novel HA epitopes. Our review also considers the role of MBCs in the effects of early-life imprinting on HA head- and stalk-specific antibody responses to influenza infection. An understanding of the processes described in this review will guide development of vaccination strategies that provide broadly effective protection.

Keywords: influenza A virus; infection; hemagglutinin; hemagglutin stalk; memory B cells; antibodies; germinal centers; original antigenic sin; imprinting

1. Introduction

B cell memory generated by influenza A virus (IAV) infection and vaccination consists of antibodies (Abs) and memory B cells (MBCs). Preexisting Abs against the virus's surface glycoproteins, the hemagglutinin (HA) and neuraminidase, have direct antiviral activity and provide the most effective protection against initiation or progression of infection [1,2]. If infection is not blocked or quickly terminated, MBC activation results in rapid and vigorous anamnestic Ab production that acts in concert with other forms of adaptive responses to clear infectious virus [3]. Activated MBCs also contribute to adaptation of the Ab response to novel features of the infecting virus by seeding germinal centers (GCs), where Ab-secreting cells and MBCs with increased binding affinity for IAV antigens are generated [4]. Although the induction of Abs by IAV infection and vaccination has been well-described [5], much less attention has been given to the essential role of MBCs in this process. Here, our goal is to review the Ab response to IAV infection in immune adults with an emphasis on the contribution of MBCs. We consider only IgG-expressing MBCs and focus entirely on the B cell response to the viral HA, the viral attachment protein that initiates cell infection by binding to sialylated receptors [6]. This enables us to consider the response of an MBC pool formed over many years by a series of exposures to related but different HAs through infection and vaccination. HA-intrinsic factors that influence MBC generation include epitope conservation as well as immunodominance hierarchies within the HA molecule [7]. In particular, our review highlights competition between preexisting MBCs and HA novelty as key determinants of the nature of HA-specific Ab production. Our review is primarily based on studies of human B cell responses. However, where appropriate, we incorporate findings from animal models that assist us to develop a more complete picture of processes in responding lymphoid tissues.

2. Anti-HA Antibodies: A Brief Overview

Each monomeric component of the homotrimeric HA consists of two structurally distinct domains: a membrane-distal head domain containing the receptor binding site and a membrane-proximal stalk domain [8]. Abs against the HA head that block binding of virus to host cells have the most potent virus neutralizing activity. However, anti-head Abs tend to be virus strain-specific because of the modification of antigenic sites in the head domain by ongoing antigenic drift. Abs against the conserved HA stalk protect via other mechanisms and are less potently neutralizing, but are more broadly reactive across HA variants and subtypes. Two phylogenetic groups of HA subtypes are recognized (group 1 and group 2), with anti-stalk Abs typically cross-reactive within a group. However, across-group stalk-binding Abs have been identified [1,9].

Abs against the IAV HA are also the basis of "original antigenic sin (OAS)" as originally termed [10], and more recently designated "antigenic seniority" [11]. Based on Ab titers measured by hemagglutination inhibition (HAI) assay, which detects Abs that bind near the receptor binding site on the HA head, IAV infection typically generates an Ab response that is broadly HA cross-reactive. Frequently, this response is of the OAS type and characterized by preferential boosting of Abs against sets of HAs related to those of viruses that circulated early in an individual's life [12,13]. The term OAS has also been applied to baseline circulating HA-reactive Ab levels that are highest against "older" HAs, a pattern that is largely maintained by OAS responses to IAV infection. It is postulated that OAS reflects the lasting imprint of early-life HA exposure, probably in the form of significant IAV infection, on an individual's immune memory. Mechanistic details of this so-called "imprinting" remain unclear, but patterns of expansion of HA-reactive MBCs are likely to be of central importance as discussed later [14,15]. In addition to imprinting effects on HA-reactive Ab production after IAV infection, the response generally includes adaptation to novel HA features and production of Abs with increased affinity for variant head epitopes [12,13,16]. The HA head domain is immunodominant over the stalk domain and is the target of the vast bulk of HA-reactive Abs produced by IAV infection. However, Abs against the stalk are also generated and, in some situations, can form a strong component of the Ab response to the HA [17].

3. MBCs and the Anti-HA Antibody Response to Influenza Infection

3.1. MBCs and Initial Antigen Encounter

B cell activation and the production of Abs against the HA of an infecting IAV begins with transport of the HA (as well as other viral antigens) from the site of infection to local lymphoid tissues and lymph nodes. Antigen binding via the B cell receptor (BCR) and uptake, processing, and presentation to establish cognate interactions with CD4 T cells are essential for MBC and naïve B cell activation [3,18]. This sets the stage for competition for antigen between different types of HA-reactive B cells, for example, between MBCs and naïve B cells, and between MBCs specific for different epitopes on the HA head and stalk. HA-reactive naïve B cells are relatively rare and are vastly outnumbered by HA-reactive MBCs in most adults. After formation, MBCs generally disperse to lymphoid tissues throughout the body [19]. However, MBCs generated by IAV infection are likely to maintain highest frequencies at sites of formation in the respiratory tract [19–21]. In addition, localization of MBCs around the periphery of lymphoid tissues or clustered near GC-like structures may allow earlier and more efficient antigen capture and CD4 T cell engagement by MBCs than by the follicle-associated naïve B cells [22,23]. MBCs are also more readily activated than naïve B cells because of cell-intrinsic factors that include epigenetic modifications, altered transcriptional networks, and greater signaling capacity of the IgG BCR [24]. In many cases, it is likely that the Ab response generated by MBC activation after IAV infection is sufficient to terminate infection with little, if any, contribution by naive B cells. Figure 1 provides a diagrammatic overview of processes described in this review.

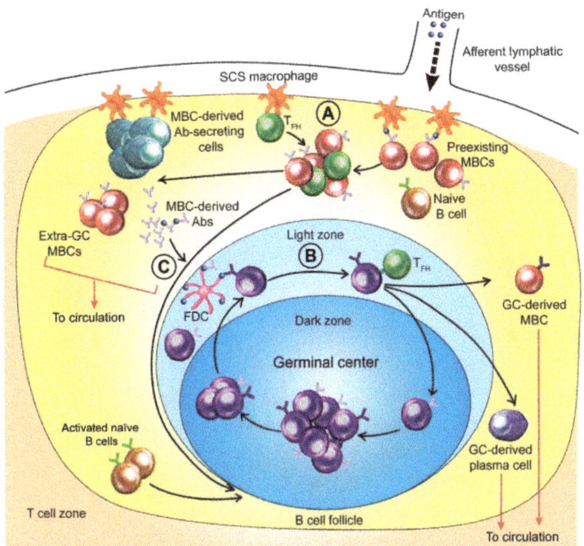

Figure 1. The HA-specific B cell response to seasonal IAV infection in an adult with a typical history of influenza exposure. Results from human and animal studies were used to construct a model of events in a responding lymph node. (**A**) Events in subcapsular proliferative foci (SPF). Influenza proteins (including HA) entering the subcapsular sinus (SCS) are trapped by SCS macrophages. HA-reactive memory B cells (MBCs) beneath the SCS compete for antigen displayed by SCS macrophages. Successful acquisition of antigen enables MBCs to receive cognate T cell help and differentiate along pathways that reflect antigen affinity: high affinity antigen binding (and stronger T cell help) favors formation of Ab-secreting cells; lower affinity cells generate MBCs or seed germinal centers (GCs). (**B**) Events in GCs. GCs are seeded by activated MBCs or activated naïve B cells. After mutation of immunoglobulin V-region genes, GC B cells that express high affinity receptors are positively selected, a process that involves competition for antigen held by follicular dendritic cells (FDCs) so that sufficiently strong T cell help can be secured. Selected cells repeat the mutation/selection cycle or differentiate into MBCs or Ab-secreting plasma cells and exit the GC. (**C**) Regulation by MBC-derived Abs. High affinity HA-reactive MBCs differentiate into Ab-secreting cells in SPF. Secreted Abs potentially regulate anti-HA Ab production by facilitating HA removal or by epitope masking. We suggest that anti-HA Abs released in the SPF enter nearby GCs and bind antigen held by FDCs. Since the initial wave of MBC-derived anti-HA Abs generally binds conserved epitopes with relatively high affinity, masking epitopes on FDC-held antigen would drive selection of GC B cells reactive to the more variant epitopes. Circulating Abs could act in a similar way, but perhaps only after development of lung inflammation increases Ab movement from the vasculature into lung tissue and then to lymph nodes via lymphatic vessels.

3.2. MBC Activation and Early Anti-HA Antibody Production

Analysis of the Ab response to IAV infection has generally focused on measurement of circulating Abs against the HA of a strain likely to represent the infecting virus, such as the strain in the most recent seasonal influenza vaccine. However, recent studies have emphasized that anti-HA Ab production is characterized by increased titers against a chronological range of HA variants [12,13,16]. Within each individual, the kinetics of the Ab response to each HA variant follows similar kinetics, indicating Ab production at a similar stage of the B cell response. These Abs are first detected in the circulation 4–6 days after symptom onset and increase to peak levels over the next seven days. Ab concentrations against each HA variant form a hierarchy, often reflecting OAS that is largely maintained over the course of the response [13]. Some studies have reported that Ab production is strongest against HAs of the infecting strain and closely related contemporary viruses [12,16]. However, this is the picture only when the response is represented as fold change in Ab titer. The pre-infection level of Abs against the HA of the infecting virus is often low and a relatively modest response to this HA can generate a marked fold increase in Ab levels. Representing the Ab response as the amount of Ab produced (delta for circulating Ab levels) results in a different picture, with strongest Ab production against older HAs [13].

Early production of anti-HA Abs following IAV infection in immune adults results from activation of preexisting MBCs, a model based on analysis of Ab-secreting plasmablasts (PBs) that appear in the circulation at about the same time as the first virus-induced Abs. These PBs generally reach peak frequencies 4–6 days post-symptom onset, but remain detectable for another 1–2 weeks [13,25]. More than 50% of total IgG PBs induced by IAV infection are virus-specific and, of these, up to 50% bind the viral HA [26]. Analysis of the HA-reactive PBs has identified recently proliferated clonal lineages that express Abs with highly mutated immunoglobulin variable genes, indicating derivation from recently activated MBCs that had previously undergone affinity maturation [26].

The site of early MBC-derived PB formation in responding lymphoid tissue has often been imprecisely described as "extrafollicular" [27]. Recently, Moran et al. [28] described a scenario for MBC activation in a skin draining lymph node that might also apply to lung-draining lymph nodes responding to IAV infection. In this scenario, resting MBCs preferentially accumulate in the outer lymph node cortex immediately beneath the subcapsular sinus (SCS) where they are in close association with SCS macrophages that span the floor of the SCS. Large numbers of memory T follicular helper (Tfh) cells occupy the same niche as the MBCs. SCS macrophages efficiently trap antigen carried by draining lymphatics to the SCS and display intact and processed antigen for recognition by memory B and T cells, respectively. Moran et al. [28] describe a structure, designated subcapsular proliferative foci (SPF) that covers the cortical surface of the B cell follicle where it abuts the SCS. Here, SCS macrophages, MBCs, and Tfh cells interact after antigen exposure, resulting in MBC proliferation and differentiation into a cluster of predominately Ab-secreting plasma cells [28,29]. This process likely generates the first HA-reactive IgG Abs produced following IAV infection, as well as the transient wave of circulating HA-reactive PBs. It follows that the SPF is a key site for competition between MBCs.

It is generally accepted that OAS patterns of HA-reactive Ab production after IAV infection reflect the competitive dominance of MBCs reactive to conserved epitopes on the HA head domain, although experimental support for this idea is lacking [14,15]. The MBC pool is established by previous HA encounter, and perhaps most importantly by the stamp of significant early-life infection on a naïve or relatively naïve B cell repertoire. In that situation, the HA-specific B cell response and MBC formation reflects the frequencies of reactive B cells in the unselected repertoire and the immunodominance hierarchy of B cell epitopes on the HA molecule. The severity/duration of early-life IAV infection is also likely to be a key factor in establishing OAS, since that would relate to the establishment and maintenance of GC reactions that expand MBCs and support affinity maturation. A series of subsequent infections by progressively more drifted IAVs would reinforce early imprinting of the B cell repertoire. This process would involve (i) preferential activation (with Ab production and MBC generation) of the more numerous, high affinity MBCs responsive to conserved HA head epitopes and (ii) the potential of anamnestic Ab production to limit infection, thus minimizing adaptation of the B cell response to novel features of the HA of the infecting virus.

The reactivity profiles of Abs expressed by individual PBs generated early in the response to IAV infection have not been extensively characterized. However, limited studies indicate that at least some of the PB-expressed Abs bind to one or more older HAs, sometimes with higher affinity than to the infecting virus HA [26,30]. These findings are consistent with PB derivation from MBCs that were generated by exposure to older HAs and respond to sufficiently conserved epitopes on the HA of the infecting virus. High affinity MBCs would be expected to outcompete lower affinity MBCs for available antigen and thus obtain the strong level of T cell help associated with differentiation of activated B cells into Ab-secreting PBs. However, it is unclear to what extent antigen affinity or simply the number of MBCs reactive to particular epitopes determines the pattern of early broad HA-reactive Ab production. The importance of MBC affinity might be diminished in the context of IAV infection, which is associated with increased levels of cytokines and Toll-like receptor agonists that help drive B cell activation [27]. Local production of soluble B cell activating factors in the subcapsular niche occupied by MBCs might be particularly important [29]. Figure 2 shows a representation of a preexisting HA-reactive MBC pool that responds to IAV infection; outcomes of MBC activation and their relationship to HA affinity are also shown. Pathways of MBC formation that we suggest maintain and/or modify the preexisting MBC pool are included in Figure 2 and are discussed later.

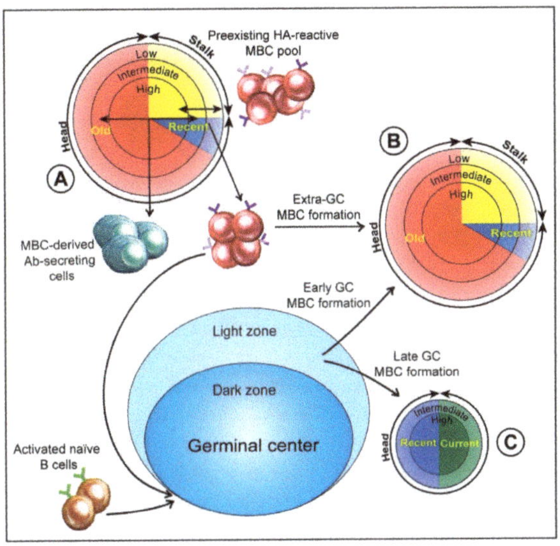

Figure 2. Composition of the MBC pool reactive to the HA of a seasonal IAV at different stages of the response to infection. MBC pools are represented as pie diagrams, with segments representing the proportions of MBCs reactive to HA head domains of older circulating IAVs ("old"; colored red), more recent IAVs ("recent"; blue), and the current circulating IAV ("current"; green), and to the stalk domain (yellow). Concentric circles identify the proportions of MBCs with high (center), intermediate, or low affinity. Composition of the preexisting HA-reactive MBC pool is based on an analysis of the response to a seasonal H3N2 IAV infection [13]; the pathways of formation and the composition of MBC pools generated during infection are speculations based on multiple studies (see Sections 3.5 and 3.6). (**A**) The preexisting MBC pool reactive to the HA of the infecting virus. Activation and differentiation of cells in this pool via extra-GC pathways generates Ab-secreting cells from high affinity MBCs and MBCs from lower affinity precursors. Activated lower affinity MBCs also seed GCs. (**B**) The MBC pool generated via the extra-GC pathway and early in the GC reaction. Weisel et al. [31] identified an early phase of GC activity that generates primarily MBCs after fewer rounds of mutation and selection. We suggest that this phase, together with the extra-GC pathway, largely reestablishes the preexisting MBC pool. (**C**) MBCs generated by prolonged GC reactions. Although there is evidence that GCs undergo a switch from MBC to plasma cell formation [31], recent studies [32,33] demonstrate that GCs can be long maintained and that the process of MBC adaptation to novel features of an HA is ongoing. We suggest that MBC-derived anti-HA Abs generated early in the response, and perhaps also circulating anti-HA Abs at a later stage, have an important regulatory role and drive positive selection of GC B cells that bind variant HA epitopes (see Figure 1C).

3.3. MBCs and Anti-HA Stalk Antibody Production

In addition to the broad HA head-reactive Ab response, approximately 10%–20% of anti-HA Abs generated by seasonal IAV infection bind to the stalk domain. The response to the stalk follows the same kinetics as the anti-head response and includes early induction of stalk-specific PBs with high numbers of mutations in immunoglobulin variable genes, indicating formation from activated MBCs [13,26]. A key determinant of the magnitude of the Ab response to the stalk is the dominance of head-reactive over stalk-reactive MBCs in competition for antigen. An individual's exposure to HA through vaccination and infection would be expected to progressively increase the ratio of head- versus stalk-reactive MBCs, primarily because of immunodominance of the head. In addition, the relatively small year-by-year changes in the HA head through antigenic drift would enable recruitment and expansion of similar head-reactive MBCs over many HA exposures. HA stalk-reactive MBCs are expanded by infection in parallel with head-reactive MBC [13]. However, evidence that stalk-binding Abs display a level of polyreactivity raises the possibility that stalk-reactive MBCs might be selected against by tolerance mechanisms [34]. Notably, few stalk-reactive Abs are produced after seasonal inactivated influenza vaccination [26,30], suggesting that factors associated with IAV infection facilitate stalk-reactive MBC activation. These might include the amount and form of the HA presented to MBCs, together with a B cell activating environment associated with infection that reduces the requirement for T cell help [27].

Studies of responses to infection and vaccination with the 2009 pandemic H1N1 (pH1N1) IAV emphasize the importance of competition between MBCs in determining the nature of HA-specific Ab production. The H1 of the pH1N1 IAV was substantially novel to most individuals less than approximately 60 years of age when the virus began circulating in humans. Initial exposure of these individuals to the pH1N1 virus through infection or vaccination generated anti-HA Abs that were broadly cross-reactive with head and stalk epitopes of multiple influenza strains [26,35,36]. In particular, the strong anti-stalk response to pH1N1 vaccination contrasted with the almost entirely head-directed response after seasonal influenza vaccination. It is now appreciated that strong anti-stalk Ab production after pH1N1 infection or vaccination reflects the novelty of the HA head domain [35]. As a result, numbers (and/or affinities) of preexisting MBCs reactive with the pH1 head are reduced, competition with stalk-reactive MBCs is lessened, and the anti-stalk Ab response is increased. The scenario is very similar when individuals are infected or vaccinated with avian IAVs, such as the H5N1 and H7N9 subtypes, which express head domains that are highly novel to most humans and stalk domains that are strongly conserved with those of seasonal IAVs. Responses to novel avian IAV HAs in adults are characterized by strong production of anti-stalk Abs, reflecting activation of preexisting stalk-reactive MBCs largely in the absence of competition from head-reactive MBCs [37,38]. Notably, the anti-HA response to avian H7N9 IAV infection consists of an early phase of broadly reactive anti-stalk Ab production, followed later by production of HAI-mediating Abs against the H7 head [39]. This pattern likely reflects early stalk-reactive MBC activation, then formation and maintenance of GC reactions that support affinity maturation of head-reactive B cells. These could include naïve B cells recruited into the response and low affinity MBCs that were not driven to PB formation. Apparently, the potent immune stimulus provided by avian IAV infection increases GC output, resulting (eventually) in production of efficiently neutralizing anti-HA head Abs that are not generated by a single dose of avian influenza vaccine [40].

3.4. MBCs and Imprinting of Antibody Responses to the HA Stalk

An effect of early-life imprinting on the character of an individual's Ab response to the HA head following IAV infection is well-established [13], but it is less clear whether responses to the stalk are also impacted. A possible scenario is that the first influenza subtype (H1N1 or H3N2) responsible for significant early-life infection imprints (through MBC expansion) not only a pattern of responsiveness to HA head epitopes, but also the potential to respond strongly to group 1 (e.g., H1) or group 2 (e.g., H3) HA stalk domains. In support of this is evidence that stalk-specific B cell lineages are established early in life and are determined in part by the subtype of the first infecting IAV [41]. Patterns of anti-stalk Ab production consistent with imprinting are unlikely to be evident in the response to seasonal IAV infection. This is because of the competitive dominance of HA head-reactive MBCs and the major role of anti-head Abs in determining the course of infection) [42]. However, the situation is very different when the HA of the infecting virus is highly novel and early anti-HA Ab production is essentially dependent on activation of preexisting MBCs reactive to the conserved stalk. This is the situation following avian IAV infection, where recovery might depend on the magnitude of anti-stalk production. Notably, Gostic et al. [43] provided evidence that individuals imprinted by a first infection with an H1N1 (group 1 HA) or an H3N2 (group 2 HA) IAV have a less severe disease when infected as adults with the avian IAVs H5N1 (group 1 HA) or H7N9 (group 2 HA), respectively. Possible explanations are higher baseline levels of circulating anti-stalk Abs or larger MBC populations reactive to the stalk of the avian HA. Recently, Tesini et al. [13] provided evidence that group 2 HA stalk-specific MBC populations expanded by seasonal H3N2 (group 2 HA) infection were largest in individuals imprinted by early-life H3N2 infection; the same was not true for group 2 HA stalk-specific Ab levels. This suggests that the imprinting effects described by Gostic et al. [43] are mediated by expanded stalk-reactive MBC populations that cross-react with either the group 1 or group 2 HA stalks of heterosubtypic IAVs.

3.5. Extra-Germinal Center Generation of MBCs

The pathways that maintain or expand HA-reactive MBC populations after IAV infection are not fully understood. For instance, it is unclear whether preexisting IgG MBCs activated in the SPF generate both PBs and MBCs via extra-GC pathways or whether MBCs are only generated in GCs after some level of remodeling of antigen reactivity. In the absence of GCs, naïve B cell activation can produce IgG MBCs that have not undergone affinity maturation [44,45], raising the possibility of an extra-GC pathway of MBC formation that maintains/expands IgG MBC populations activated to generate the early PB response (see Figure 2). In line with this idea is the observation that recently proliferated, HA-reactive MBCs increase in the circulation 4–6 days after onset of symptoms post-IAV infection, closely accompanying the initial increase in HA-reactive PBs and Ab levels [13,46]. Notably, the recent analysis by Tesini et al. [13] demonstrated a close relationship between patterns of early Ab production after IAV infection measured against a range of HA variants (including OAS patterns) and patterns of early expansion of MBC populations reactive to the same set of HAs. This result is consistent with Ab production and MBC formation stemming from activation of the same precursor MBC. B cell clonal lineage analysis by Ellebedy et al. [46] also indicated that PB and MBC lineages could have originated from the same activated MBC, but does not exclude a contribution of GCs to at least part of the MBC lineage.

3.6. Germinal Center Events

A subset of HA-reactive B cells activated by IAV infection enter GC reactions where they undergo affinity maturation [47]. A proportion of the preexisting MBCs activated by infection become GC B cells as an alternative to extra-GC differentiation into PBs or MBCs. Activated naïve B cells might also enter GCs, especially when infection is severe and the response is prolonged [26]. A key determinant of the fate decision of an activated naïve or memory B cell is affinity for antigen, with high affinity linking directly to strong T cell help and favoring Ab-secreting cell formation [48,49]. Entry of activated cells into GCs was associated with an "intermediate" level of T cell help [50]. However, the threshold affinity requirement for GC seeding could be modulated by increased levels of B cell activating factors in the infection-associated environment, resulting in GC entry of low affinity B cells that might otherwise have formed MBCs or been lost from the response [3,27,51,52].

After Ab V-region somatic hypermutation, GC B cells that express BCRs with high antigen affinity are positively selected for further cell division and mutation or for differentiation along pathways that generate plasma cells or MBCs. The process of positive selection is not fully understood, but signaling to GC B cells via the BCR and CD40 plays a central role. A key step is competition between GC B cells for antigen displayed on the surface of follicular dendritic cells (FDCs); Ag that is bound with sufficient affinity by GC B cells is taken up, processed, and presented to secure cognate help from Tfh cells [53]. Importantly, Tas et al. [54] provided evidence that GCs can be seeded by a highly oligoclonal pool of activated B cells (estimated at 10 s to 100 s of clones per GC). This sets the stage for a combination of two types of competition between GC B cells within single GCs: intraclonal competition between mutational variants of the same clone (specific for the same epitope), and interclonal competition between clones specific for different epitopes on the same or different antigens. In addition to numbers and affinities of specific GC B cells and antigen abundance and form on FDCs, antigen-intrinsic immunodominance hierarchies might also play an important role in determining the outcome of interclonal competition in GCs that (for example) drive selection of HA head-reactive over stalk-reactive GC B cells.

Recent studies of human responses to avian HAs, including H5 expressed by a replicating adenovirus, provide evidence that GC activity continues for many months after immunization [32,33]. It is, therefore, reasonable to expect long-lived GC reactions after seasonal IAV infection, allowing factors such as affinity-matured Abs or a general reduction in available antigen to continue to influence selection of GC B cells and drive affinity maturation. The signals that direct selected GC B cells along the plasma cell or MBC differentiation pathways before exiting the GC are similar to those responsible for extra-GC fate decisions. Generally, high affinity GC B cells that acquire antigen efficiently and receive strong T cell help form plasma cells; lower affinity B cells enter the MBC pool [47]. Experiments in mice indicate that GCs initially generate mostly MBCs and then switch after two or more weeks to plasma cell formation [31]. MBC formation after fewer rounds of selection would limit the extent of adaptation to the HA of the infecting virus, but might provide the advantage of expanding MBCs with greater breadth of reactivity. The temporal switch in GC output from MBCs to plasma cells fits with ongoing cycles of mutation and selection and the preferential formation of plasma cells from B cells expressing high affinity receptors. There is nevertheless a degree of affinity maturation of MBCs generated in GCs. Tesini et al. [13] analyzed responses to seasonal H3N2 infection in the 2012–2013 season when infecting viruses were HA drift variants [55]. Circulating MBCs adapted to the H3 head domain and perhaps generated during the early phase of MBC production in GCs were detected within approximately four weeks of symptom onset. However, evidence that HA-reactive MBCs with increased levels of somatic hypermutation are formed many months after HA exposure indicates that affinity maturation of MBCs continues long after a switch by GCs to predominantly plasma cell production (see Figure 2) [32,33].

3.7. Regulation of the Anti-HA Response by MBC-Derived Antibodies

Preexisting anti-HA Abs and anti-HA Abs generated at various stages after IAV infection are thought to play an important role in regulating the fine specificity of the anti-HA Ab response [15]. The anti-HA Abs could block B cell recognition of particular epitopes/antigenic sites by epitope masking or more generally suppress the response by facilitating HA removal through Fc-mediated mechanisms [56,57]. Epitope masking by preexisting Abs does not fit with the observation that patterns of HA-reactive Ab production early in the response to IAV infection directly reflect OAS hierarchies of HA-reactive Ab levels in the circulation prior to infection [13]. An important consideration is whether or when preexisting circulating Abs have sufficient access to sites of immune response generation to exert a regulatory influence. For instance, circulating Abs might always have ready access to events in the spleen, but might not reach high levels in lymph nodes responding to IAV infection until anti-viral responses commence in the respiratory tract and extend to draining lymph nodes [58,59]. We suggest that preexisting circulating Abs have little if any effect on MBC activation in the SPF of local lymph nodes; this component of the response generates the first anti-HA Abs and is primarily regulated by MBC competition. Furthermore, we speculate that Abs resulting from early MBC activation in the SPF play a key role in response regulation by mechanisms like epitope masking. Perhaps most important is the potential for these Abs to regulate positive selection in GCs, a possibility suggested by the close proximity of the SPF and B cell follicles. Free Abs might enter

GCs and interact with antigen on FDC networks, or bind antigen outside GCs and enter as immune complexes. Zhang et al. [60] demonstrated that Abs enter GCs, bind antigen held on FDCs, and compete with GC B cells in an affinity-dependent way. Through these mechanisms, Abs derived from preexisting MBCs could facilitate affinity maturation of responses to drifted epitopes on the HA head [61]. In the case of MBC-derived anti-stalk Abs, GC B cell selection would be shifted towards targeting the immunodominant HA head [62].

4. Summary

Our goal in this review is to combine observations from human and animal studies to provide a picture of the HA-specific B cell response to IAV infection in adults. A critical determinant of the nature of the response in adults is the HA-specific B cell memory established by HA exposure through IAV infection and vaccination over many years. Of central importance is the composition of an individual's HA-reactive MBC pool, a population that even in adulthood reflects the imprint of early-life HA exposure and is responsible for OAS patterns of Ab production early in the response to IAV infection. PB formation from activated preexisting MBCs with high affinity for the HA of the infecting virus optimizes the protective efficacy of the initial wave of anti-HA Abs. We emphasize MBC competition as a key determinant of the patterns of activation of preexisting MBCs. The novelty of the HA of the infecting virus has a profound influence on the outcome of MBC competition and (for example) determines the strength of Ab production against the HA stalk. Preexisting MBC activation generates cells that along with activated naïve B cells, seed GCs for generation of MBCs and plasma cells adapted to variant/novel HA epitopes. This process involves another form of B cell competition, the competition between GC B cells for antigen held by FDCs; acquisition of this antigen is a prerequisite for positive selection. We suggest that Abs resulting from preexisting MBC activation, more so than preexisting circulating Abs, play an important regulatory role in positive selection through epitope masking. Depending on the epitope specificity of Abs derived from preexisting MBCs, GC B cells will be refocused on the immunodominant HA head or on particular drifted head epitopes. Our review also considers the generation and expansion of MBCs at different points in the response to both maintain OAS response patterns and adapt to emerging HAs. Our increasing understanding of MBC competition, immunodominance hierarchies, and Ab regulation of B cell responses will continue to guide development of vaccine composition and administration strategies to optimize B cell-mediated protection generated against IAV infection [63–67].

Author Contributions: Writing—original draft preparation, M.Y.S.; writing—review and editing, M.Y.S., P.Q.T.N. and D.J.T.; visualization, M.Y.S. and P.Q.T.N.

Funding: This work was supported with Federal funds from the National Institute of Allergy and Infectious Diseases, NIH, Department of Health and Human Services through an award to the New York Influenza Center of Excellence (NYICE, Contract No. HHSN272201400005C), a member of the NIAID Centers of Excellence for Influenza Research and Surveillance (CEIRS).

Conflicts of Interest: The authors declare no conflict of interest.

References

1. Krammer, F.; Palese, P. Influenza virus hemagglutinin stalk-based antibodies and vaccines. *Curr. Opin. Virol.* **2013**, *3*, 521–530. [CrossRef] [PubMed]
2. Krammer, F.; Fouchier, R.A.M.; Eichelberger, M.C.; Webby, R.J.; Shaw-Saliba, K.; Wan, H.; Wilson, P.C.; Compans, R.W.; Skountzou, I.; Monto, A.S. NAction! How can neuraminidase-based immunity contribute to better influenza virus vaccines? *MBio* **2018**, *9*. [CrossRef] [PubMed]
3. Cyster, J.G.; Allen, C.D.C. B cell responses: Cell interaction dynamics and decisions. *Cell* **2019**, *177*, 524–540. [CrossRef]
4. Shlomchik, M.J. Do memory B cells form secondary germinal centers? Yes and no. *Cold Spring Harb. Perspect Biol.* **2018**, *10*, a029405. [CrossRef] [PubMed]

5. Krammer, F. The human antibody response to influenza A virus infection and vaccination. *Nat. Rev. Immunol.* **2019**, *19*, 383–397. [CrossRef] [PubMed]
6. Russell, C.J.; Hu, M.; Okda, F.A. Influenza hemagglutinin protein stability, activation, and pandemic risk. *Trends Microbiol.* **2018**, *26*, 841–853. [CrossRef] [PubMed]
7. Zost, S.J.; Wu, N.C.; Hensley, S.E.; Wilson, I.A. Immunodominance and antigenic variation of influenza virus hemagglutinin: Implications for design of universal vaccine immunogens. *J. Infect. Dis.* **2019**, *219*, S38–S45. [CrossRef] [PubMed]
8. Wu, N.C.; Wilson, I.A. Structural insights into the design of novel anti-influenza therapies. *Nat. Struct. Mol. Biol.* **2018**, *25*, 115–121. [CrossRef] [PubMed]
9. Ekiert, D.C.; Wilson, I.A. Broadly neutralizing antibodies against influenza virus and prospects for universal therapies. *Curr. Opin. Virol.* **2012**, *2*, 134–141. [CrossRef]
10. Francis, T. On the doctrine of original antigenic sin. *Proc. Am. Philos. Soc.* **1960**, *104*, 572–578.
11. Lessler, J.; Riley, S.; Read, J.M.; Wang, S.; Zhu, H.; Smith, G.J.D.; Guan, Y.; Jiang, C.Q.; Cummings, D.A.T. Evidence for antigenic seniority in influenza A (H3N2) antibody responses in southern China. *PLoS Pathog.* **2012**, *8*, e1002802. [CrossRef] [PubMed]
12. Fonville, J.M.; Wilks, S.H.; James, S.L.; Fox, A.; Ventresca, M.; Aban, M.; Xue, L.; Jones, T.C.; Le, N.M.H.; Pham, Q.T.; et al. Antibody landscapes after influenza virus infection or vaccination. *Science* **2014**, *346*, 996–1000. [CrossRef] [PubMed]
13. Tesini, B.L.; Kanagaiah, P.; Wang, J.; Hahn, M.; Halliley, J.L.; Chaves, F.A.; Nguyen, P.Q.T.; Nogales, A.; DeDiego, M.L.; Anderson, C.S.; et al. Broad hemagglutinin-specific memory B Cell expansion by seasonal influenza virus infection reflects early-life imprinting and adaptation to the infecting virus. *J. Virol.* **2019**, *93*, e00169-19. [CrossRef] [PubMed]
14. Cobey, S.; Hensley, S.E. Immune history and influenza virus susceptibility. *Curr. Opin. Virol.* **2017**, *22*, 105–111. [CrossRef] [PubMed]
15. Henry, C.; Palm, A.-K.E.; Krammer, F.; Wilson, P.C. From original antigenic sin to the universal influenza virus vaccine. *Trends Immunol.* **2018**, *39*, 70–79. [CrossRef] [PubMed]
16. Nachbagauer, R.; Choi, A.; Hirsh, A.; Margine, I.; Iida, S.; Barrera, A.; Ferres, M.; Albrecht, R.A.; García-Sastre, A.; Bouvier, N.M.; et al. Defining the antibody cross-reactome directed against the influenza virus surface glycoproteins. *Nat. Immunol.* **2017**, *18*, 464–473. [CrossRef]
17. Neu, K.E.; Henry Dunand, C.J.; Wilson, P.C. Heads, stalks and everything else: How can antibodies eradicate influenza as a human disease? *Curr. Opin. Immunol.* **2016**, *42*, 48–55. [CrossRef]
18. Cyster, J.G. B cell follicles and antigen encounters of the third kind. *Nat. Immunol.* **2010**, *11*, 989–996. [CrossRef]
19. Joo, H.M.; He, Y.; Sangster, M.Y. Broad dispersion and lung localization of virus-specific memory B cells induced by influenza pneumonia. *Proc. Natl. Acad. Sci. USA* **2008**, *105*, 3485–3490. [CrossRef]
20. Pichyangkul, S.; Yongvanitchit, K.; Limsalakpetch, A.; Kum-Arb, U.; Im-Erbsin, R.; Boonnak, K.; Thitithayanont, A.; Jongkaewwattana, A.; Wiboon-ut, S.; Mongkolsirichaikul, D.; et al. Tissue distribution of memory T and B Cells in rhesus monkeys following influenza A infection. *J. Immunol.* **2015**, *195*, 4378–4386. [CrossRef]
21. Jegaskanda, S.; Mason, R.D.; Andrews, S.F.; Wheatley, A.K.; Zhang, R.; Reynoso, G.V.; Ambrozak, D.R.; Santos, C.P.; Luke, C.J.; Matsuoka, Y.; et al. Intranasal live influenza vaccine priming elicits localized B cell responses in mediastinal lymph nodes. *J. Virol.* **2018**, *92*, e01970-17. [CrossRef] [PubMed]
22. Aiba, Y.; Kometani, K.; Hamadate, M.; Moriyama, S.; Sakaue-Sawano, A.; Tomura, M.; Luche, H.; Fehling, H.J.; Casellas, R.; Kanagawa, O.; et al. Preferential localization of IgG memory B cells adjacent to contracted germinal centers. *Proc. Natl. Acad. Sci. USA* **2010**, *107*, 12192–12197. [CrossRef] [PubMed]
23. Zhao, Y.; Uduman, M.; Siu, J.H.Y.; Tull, T.J.; Sanderson, J.D.; Wu, Y.-C.B.; Zhou, J.Q.; Petrov, N.; Ellis, R.; Todd, K.; et al. Spatiotemporal segregation of human marginal zone and memory B cell populations in lymphoid tissue. *Nat. Commun.* **2018**, *9*, 3857. [CrossRef]
24. Good-Jacobson, K.L. Strength in diversity: Phenotypic, functional, and molecular heterogeneity within the memory B cell repertoire. *Immunol. Rev.* **2018**, *284*, 67–78. [CrossRef] [PubMed]
25. Huang, K.-Y.A.; Li, C.K.-F.; Clutterbuck, E.; Chui, C.; Wilkinson, T.; Gilbert, A.; Oxford, J.; Lambkin-Williams, R.; Lin, T.-Y.; McMichael, A.J.; et al. Virus-specific antibody secreting cell, memory B-cell, and sero-antibody responses in the human influenza challenge model. *J. Infect. Dis.* **2014**, *209*, 1354–1361. [CrossRef]

26. Wrammert, J.; Koutsonanos, D.; Li, G.-M.; Edupuganti, S.; Sui, J.; Morrissey, M.; McCausland, M.; Skountzou, I.; Hornig, M.; Lipkin, W.I.; et al. Broadly cross-reactive antibodies dominate the human B cell response against 2009 pandemic H1N1 influenza virus infection. *J. Exp. Med.* **2011**, *208*, 181–193. [CrossRef] [PubMed]
27. Lam, J.H.; Baumgarth, N. The multifaceted B cell response to influenza virus. *J. Immunol.* **2019**, *202*, 351–359. [CrossRef]
28. Moran, I.; Nguyen, A.; Khoo, W.H.; Butt, D.; Bourne, K.; Young, C.; Hermes, J.R.; Biro, M.; Gracie, G.; Ma, C.S.; et al. Memory B cells are reactivated in subcapsular proliferative foci of lymph nodes. *Nat. Commun.* **2018**, *9*, 3372. [CrossRef]
29. Moran, I.; Grootveld, A.K.; Nguyen, A.; Phan, T.G. Subcapsular sinus macrophages: The seat of innate and adaptive memory in murine lymph nodes. *Trends Immunol.* **2019**, *40*, 35–48. [CrossRef]
30. Moody, M.A.; Zhang, R.; Walter, E.B.; Woods, C.W.; Ginsburg, G.S.; McClain, M.T.; Denny, T.N.; Chen, X.; Munshaw, S.; Marshall, D.J.; et al. H3N2 influenza infection elicits more cross-reactive and less clonally expanded anti-hemagglutinin antibodies than influenza vaccination. *PLoS ONE* **2011**, *6*, e25797. [CrossRef]
31. Weisel, F.J.; Zuccarino-Catania, G.V.; Chikina, M.; Shlomchik, M.J. A temporal switch in the germinal center determines differential output of memory B and plasma cells. *Immunity* **2016**, *44*, 116–130. [CrossRef] [PubMed]
32. Andrews, S.F.; Chambers, M.J.; Schramm, C.A.; Plyler, J.; Raab, J.E.; Kanekiyo, M.; Gillespie, R.A.; Ransier, A.; Darko, S.; Hu, J.; et al. Activation dynamics and immunoglobulin evolution of pre-existing and newly generated human memory B cell responses to influenza hemagglutinin. *Immunity* **2019**. [CrossRef] [PubMed]
33. Matsuda, K.; Huang, J.; Zhou, T.; Sheng, Z.; Kang, B.H.; Ishida, E.; Griesman, T.; Stuccio, S.; Bolkhovitinov, L.; Wohlbold, T.J.; et al. Prolonged evolution of the memory B cell response induced by a replicating adenovirus-influenza H5 vaccine. *Sci. Immunol.* **2019**, *4*, eaau2710. [CrossRef] [PubMed]
34. Andrews, S.F.; Huang, Y.; Kaur, K.; Popova, L.I.; Ho, I.Y.; Pauli, N.T.; Dunand, C.J.H.; Taylor, W.M.; Lim, S.; Huang, M.; et al. Immune history profoundly affects broadly protective B cell responses to influenza. *Sci. Transl. Med.* **2015**, *7*, 316ra192. [CrossRef]
35. Li, G.-M.; Chiu, C.; Wrammert, J.; McCausland, M.; Andrews, S.F.; Zheng, N.-Y.; Lee, J.-H.; Huang, M.; Qu, X.; Edupuganti, S.; et al. Pandemic H1N1 influenza vaccine induces a recall response in humans that favors broadly cross-reactive memory B cells. *Proc. Natl. Acad. Sci. USA* **2012**, *109*, 9047–9052. [CrossRef]
36. Sangster, M.Y.; Baer, J.; Santiago, F.W.; Fitzgerald, T.; Ilyushina, N.A.; Sundararajan, A.; Henn, A.D.; Krammer, F.; Yang, H.; Luke, C.J.; et al. B cell response and hemagglutinin stalk-reactive antibody production in different age cohorts following 2009 H1N1 influenza virus vaccination. *Clin. Vaccine Immunol.* **2013**, *20*, 867–876. [CrossRef] [PubMed]
37. Ellebedy, A.H.; Krammer, F.; Li, G.-M.; Miller, M.S.; Chiu, C.; Wrammert, J.; Chang, C.Y.; Davis, C.W.; McCausland, M.; Elbein, R.; et al. Induction of broadly cross-reactive antibody responses to the influenza HA stem region following H5N1 vaccination in humans. *Proc. Natl. Acad. Sci. USA* **2014**, *111*, 13133–13138. [CrossRef]
38. Stadlbauer, D.; Rajabhathor, A.; Amanat, F.; Kaplan, D.; Masud, A.; Treanor, J.J.; Izikson, R.; Cox, M.M.; Nachbagauer, R.; Krammer, F. Vaccination with a recombinant H7 hemagglutinin-based influenza virus vaccine induces broadly reactive antibodies in humans. *mSphere* **2017**, *2*, e00502–e00517. [CrossRef]
39. Liu, L.; Nachbagauer, R.; Zhu, L.; Huang, Y.; Xie, X.; Jin, S.; Zhang, A.; Wan, Y.; Hirsh, A.; Tian, D.; et al. Induction of broadly cross-reactive stalk-specific antibody responses to influenza group 1 and group 2 hemagglutinins by natural H7N9 virus infection in humans. *J. Infect. Dis.* **2017**, *215*, 518–528. [CrossRef]
40. Topham, D.J.; Nguyen, P.; Sangster, M.Y. Pandemic influenza vaccines: What they have taught us about B cell immunology. *Curr. Opin. Immunol.* **2018**, *53*, 203–208. [CrossRef]
41. Andrews, S.F.; Joyce, M.G.; Chambers, M.J.; Gillespie, R.A.; Kanekiyo, M.; Leung, K.; Yang, E.S.; Tsybovsky, Y.; Wheatley, A.K.; Crank, M.C.; et al. Preferential induction of cross-group influenza A hemagglutinin stem–specific memory B cells after H7N9 immunization in humans. *Sci. Immunol.* **2017**, *2*, eaan2676. [CrossRef] [PubMed]
42. Christensen, S.R.; Toulmin, S.A.; Griesman, T.; Lamerato, L.E.; Petrie, J.G.; Martin, E.T.; Monto, A.S.; Hensley, S.E. Assessing the protective potential of H1N1 influenza virus hemagglutinin head and stalk antibodies in humans. *J. Virol.* **2019**, *93*, e02134-18. [CrossRef] [PubMed]
43. Gostic, K.M.; Ambrose, M.; Worobey, M.; Lloyd-Smith, J.O. Potent protection against H5N1 and H7N9 influenza via childhood hemagglutinin imprinting. *Science* **2016**, *354*, 722–726. [CrossRef] [PubMed]

44. Kaji, T.; Ishige, A.; Hikida, M.; Taka, J.; Hijikata, A.; Kubo, M.; Nagashima, T.; Takahashi, Y.; Kurosaki, T.; Okada, M.; et al. Distinct cellular pathways select germline-encoded and somatically mutated antibodies into immunological memory. *J. Exp. Med.* **2012**, *209*, 2079–2097. [CrossRef] [PubMed]
45. Inamine, A.; Takahashi, Y.; Baba, N.; Miyake, K.; Tokuhisa, T.; Takemori, T.; Abe, R. Two waves of memory B-cell generation in the primary immune response. *Int. Immunol.* **2005**, *17*, 581–589. [CrossRef] [PubMed]
46. Ellebedy, A.H.; Jackson, K.J.L.; Kissick, H.T.; Nakaya, H.I.; Davis, C.W.; Roskin, K.M.; McElroy, A.K.; Oshansky, C.M.; Elbein, R.; Thomas, S.; et al. Defining antigen-specific plasmablast and memory B cell subsets in human blood after viral infection or vaccination. *Nat. Immunol.* **2016**, *17*, 1226–1234. [CrossRef] [PubMed]
47. Mesin, L.; Ersching, J.; Victora, G.D. Germinal center B cell dynamics. *Immunity* **2016**, *45*, 471–482. [CrossRef]
48. Paus, D.; Phan, T.G.; Chan, T.D.; Gardam, S.; Basten, A.; Brink, R. Antigen recognition strength regulates the choice between extrafollicular plasma cell and germinal center B cell differentiation. *J. Exp. Med.* **2006**, *203*, 1081–1091. [CrossRef]
49. Chan, T.D.; Gatto, D.; Wood, K.; Camidge, T.; Basten, A.; Brink, R. Antigen affinity controls rapid T-dependent antibody production by driving the expansion rather than the differentiation or extrafollicular migration of early plasmablasts. *J. Immunol.* **2009**, *183*, 3139–3149. [CrossRef]
50. Schwickert, T.A.; Victora, G.D.; Fooksman, D.R.; Kamphorst, A.O.; Mugnier, M.R.; Gitlin, A.D.; Dustin, M.L.; Nussenzweig, M.C. A dynamic T cell–limited checkpoint regulates affinity-dependent B cell entry into the germinal center. *J. Exp. Med.* **2011**, *208*, 1243–1252. [CrossRef]
51. Bannard, O.; Cyster, J.G. Germinal centers: Programmed for affinity maturation and antibody diversification. *Curr. Opin. Immunol.* **2017**, *45*, 21–30. [CrossRef] [PubMed]
52. Inoue, T.; Moran, I.; Shinnakasu, R.; Phan, T.G.; Kurosaki, T. Generation of memory B cells and their reactivation. *Immunol. Rev.* **2018**, *283*, 138–149. [CrossRef] [PubMed]
53. Heesters, B.A.; van der Poel, C.E.; Das, A.; Carroll, M.C. Antigen presentation to B cells. *Trends Immunol.* **2016**, *37*, 844–854. [CrossRef] [PubMed]
54. Tas, J.M.J.; Mesin, L.; Pasqual, G.; Targ, S.; Jacobsen, J.T.; Mano, Y.M.; Chen, C.S.; Weill, J.-C.; Reynaud, C.-A.; Browne, E.P.; et al. Visualizing antibody affinity maturation in germinal centers. *Science* **2016**, *351*, 1048–1054. [CrossRef] [PubMed]
55. DeDiego, M.L.; Anderson, C.S.; Yang, H.; Holden-Wiltse, J.; Fitzgerald, T.; Treanor, J.J.; Topham, D.J. Directed selection of influenza virus produces antigenic variants that match circulating human virus isolates and escape from vaccine-mediated immune protection. *Immunology* **2016**, *148*, 160–173. [CrossRef] [PubMed]
56. Zarnitsyna, V.I.; Ellebedy, A.H.; Davis, C.; Jacob, J.; Ahmed, R.; Antia, R. Masking of antigenic epitopes by antibodies shapes the humoral immune response to influenza. *Philos. Trans. R. Soc. B* **2015**, *370*, 20140248. [CrossRef] [PubMed]
57. Bergström, J.J.E.; Xu, H.; Heyman, B. Epitope-specific suppression of IgG responses by passively administered specific IgG: Evidence of epitope masking. *Front. Immunol.* **2017**, *8*, 238. [CrossRef] [PubMed]
58. Hermesh, T.; Moltedo, B.; López, C.B.; Moran, T.M. Buying time—The immune system determinants of the incubation period to respiratory viruses. *Viruses* **2010**, *2*, 2541–2558. [CrossRef] [PubMed]
59. Tabrizi, M.; Bornstein, G.G.; Suria, H. Biodistribution mechanisms of therapeutic monoclonal antibodies in health and disease. *AAPS J.* **2010**, *12*, 33–43. [CrossRef] [PubMed]
60. Zhang, Y.; Meyer-Hermann, M.; George, L.A.; Figge, M.T.; Khan, M.; Goodall, M.; Young, S.P.; Reynolds, A.; Falciani, F.; Waisman, A.; et al. Germinal center B cells govern their own fate via antibody feedback. *J. Exp. Med.* **2013**, *210*, 457–464. [CrossRef]
61. Wang, T.T.; Bournazos, S.; Ravetch, J.V. Immunological responses to influenza vaccination: Lessons for improving vaccine efficacy. *Curr. Opin. Immunol.* **2018**, *53*, 124–129. [PubMed]
62. Victora, G.D.; Wilson, P.C. Germinal center selection and the antibody response to influenza. *Cell* **2015**, *163*, 545–548. [CrossRef] [PubMed]
63. Yassine, H.M.; Boyington, J.C.; McTamney, P.M.; Wei, C.-J.; Kanekiyo, M.; Kong, W.-P.; Gallagher, J.R.; Wang, L.; Zhang, Y.; Joyce, M.G.; et al. Hemagglutinin-stem nanoparticles generate heterosubtypic influenza protection. *Nat. Med.* **2015**, *21*, 1065–1070. [CrossRef] [PubMed]
64. Broecker, F.; Liu, S.T.H.; Suntronwong, N.; Sun, W.; Bailey, M.J.; Nachbagauer, R.; Krammer, F.; Palese, P. A mosaic hemagglutinin-based influenza virus vaccine candidate protects mice from challenge with divergent H3N2 strains. *NPJ Vaccines* **2019**, *4*, 31. [CrossRef] [PubMed]

65. Kanekiyo, M.; Joyce, M.G.; Gillespie, R.A.; Gallagher, J.R.; Andrews, S.F.; Yassine, H.M.; Wheatley, A.K.; Fisher, B.E.; Ambrozak, D.R.; Creanga, A.; et al. Mosaic nanoparticle display of diverse influenza virus hemagglutinins elicits broad B cell responses. *Nat. Immunol.* **2019**, *20*, 362–372. [CrossRef]
66. Choi, A.; Bouzya, B.; Cortés Franco, K.-D.; Stadlbauer, D.; Rajabhathor, A.; Rouxel, R.N.; Mainil, R.; Van der Wielen, M.; Palese, P.; García-Sastre, A.; et al. Chimeric hemagglutinin-based influenza virus vaccines induce protective stalk-specific humoral immunity and cellular responses in mice. *ImmunoHorizons* **2019**, *3*, 133–148. [CrossRef] [PubMed]
67. Angeletti, D.; Kosik, I.; Santos, J.J.S.; Yewdell, W.T.; Boudreau, C.M.; Mallajosyula, V.V.A.; Mankowski, M.C.; Chambers, M.; Prabhakaran, M.; Hickman, H.D.; et al. Outflanking immunodominance to target subdominant broadly neutralizing epitopes. *Proc. Natl. Acad. Sci. USA* **2019**, *116*, 13474–13479. [CrossRef] [PubMed]

pathogens

MDPI

Review

The Importance of Vaccinating Children and Pregnant Women against Influenza Virus Infection

Ravi S Misra [1,*] and Jennifer L Nayak [2]

1 Department of Pediatrics Division of Neonatology, The University of Rochester Medical Center, Rochester, NY 14623, USA

2 Department of Pediatrics Division of Pediatric Infectious Diseases, The University of Rochester Medical Center, Rochester, NY 14623, USA; jennifer_nayak@urmc.rochester.edu

* Correspondence: ravi_misra@urmc.rochester.edu

Received: 25 September 2019; Accepted: 22 November 2019; Published: 26 November 2019

Abstract: Influenza virus infection is responsible for significant morbidity and mortality in the pediatric and pregnant women populations, with deaths frequently caused by severe influenza-associated lower respiratory tract infection and acute respiratory distress syndrome (ARDS). An appropriate immune response requires controlling the viral infection through activation of antiviral defenses, which involves cells of the lung and immune system. High levels of viral infection or high levels of inflammation in the lower airways can contribute to ARDS. Pregnant women and young children, especially those born prematurely, may develop serious complications if infected with influenza virus. Vaccination against influenza virus will lead to lower infection rates and fewer complications, even if the vaccine is poorly matched to circulating viral strains. Maternal vaccination offers infants protection via antibody transmission through the placenta and breast milk. Despite the health benefits of the influenza vaccine, vaccination rates around the world remain well below targets. Trust in the use of vaccines among the public must be restored in order to increase vaccination rates and decrease the public health burden of influenza.

Keywords: pediatrics; influenza virus; vaccines; pregnant women; ARDS; lung; morbidity; vaccination rate

1. Introduction

Influenza virus infection is a major cause of morbidity and mortality around the world, with over 3 million individuals developing severe disease and resulting in hundreds of thousands of deaths per year [1]. Some of the most vulnerable populations include pregnant women and young children, making these groups a high priority target for vaccination. Individuals who are less than 21 years old are considered to be in the pediatric group and subsets include neonates (0 to 28 days), infants (29 days to 2 years), children (2 years to less than 12 years), and adolescents (12 years to 21 years) [2]. For individuals who are less than 5 years of age, pediatric influenza-associated infections are estimated at 90 million cases per year, with 1 million cases of influenza-associated severe acute lower respiratory tract infection and 28,000–111,500 deaths, the majority of which occur in developing countries [3]. Children less than a year of age are particularly susceptible to infection as they have little pre-existing immunity and may be too young to be vaccinated, relying upon transferred maternal immunity to protect against infection [4–7].

Several studies report that in fatal cases of influenza virus infection, increased inflammation, and virus are found in the alveoli [8–13]. Disruption of the alveolar region due to viral infection and increased inflammation can contribute to the development of acute respiratory distress syndrome (ARDS), which is a major health concern for children and pregnant women [14–17]. Vaccination can help prevent infection, which will, in turn, prevent acute lower respiratory tract infections and

ARDS. Vaccines are either in the form of Inactivated Influenza Vaccine (IIV) or Live Attenuated Influenza Virus (LAIV) [18]. Current influenza vaccines target predominately the variable region of the hemagglutinin protein. This can allow for viruses to escape the immune system via mutation, leading to vaccine mismatch and increased viral spread [19–27]. Scientists are continuing work to develop a universal influenza vaccine that targets a less variable region of the influenza virus and is thus protective against a greater breadth of viral strains but, while substantial progress has been made, challenges remain [28–30]. This review will discuss our current understanding of the immune response to influenza vaccination, with a focus on the benefits of vaccinating pregnant women and children against influenza virus.

2. Body

2.1. Background

Influenza virus is a single stranded, negative sense Orthomyxoviridae RNA virus with a genome that contains 8 genomic segments [31]. Influenza A and B viruses include the hemagglutinin (HA), neuraminidase (NA), nucleoprotein (NP), matrix protein 1 (M1), matrix protein 2 (M2/BM2), nonstructural protein 1 (NS1), nonstructural protein 2 (NS2) and the RNA polymerase complex (PA, PB1 and PB2), as illustrated in Figure 1. The strain of influenza A virus is defined by the combination of HA and NA proteins, of which there are 18 and 11 distinct subtypes identified, respectively [31]. HA subtypes are classified as group 1 (H1, H2, H5, H6, H8, H9, H11, H12, H13, H16, H17, and H18) or group 2 (H3, H4 H7, H10, H14 and H15) [28]. While influenza A virus remains the major focus of public health officials, the pathogenic potential risk of influenza B virus infections must not be ignored [32,33]. In fact, several studies implicate influenza B viral infection as a substantial health concern in the young pediatric population [34–36]. Both molecular and host determinants, such as increased viral replication, host cell death, host antiviral gene response, degree of pre-existing immunity, and transmissibility, can contribute to the pathogenicity of influenza virus [37].

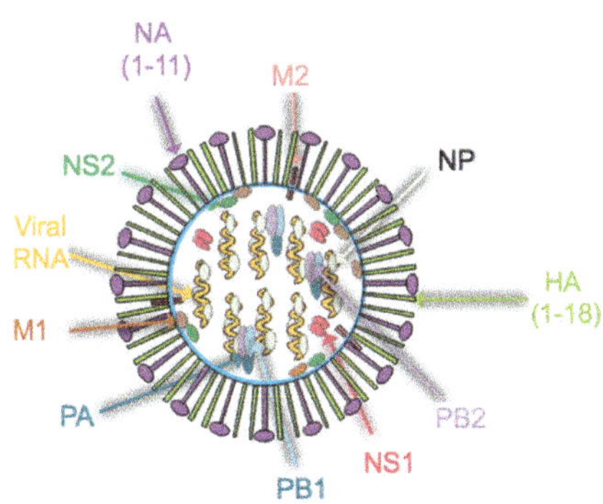

Figure 1. Diagram of the Orthomyxoviridae influenza virus. The virus consists of neuraminidase (NA), matrix protein 1 (M1), matrix 2 proton pump (M2), hemagglutinin (HA), polymerase acid subunits (PB1, PB2, PA), non-structural (NS1, NS2) and nucleoprotein (NP). Current vaccine strategies target the head regions of HA, which are highly variable between viral strains. Universal influenza vaccines target less-variable regions of the virus, such as the HA stalk, which will provide coverage for a range of influenza virus strains.

Given the relatively error-prone nature of the influenza virus RNA polymerases, mutations are often introduced which can result in viral escape from host immunity through the process of antigenic drift. Such mutations often are selected for in HA as they can lead to an escape from viral neutralization [19–23]. However, antigenic drift can also be seen in other viral proteins [24,25,27,38]. Alternatively, antigenic shift can occur when viral gene segment recombination results in formation of a novel virus against which there is little pre-existing host immunity [39,40]. As the general population is often relatively immunologically naïve to a shifted viral strain, such a virus may be able to spread rapidly across the globe potentially causing an influenza pandemic [41–44]. Given that animals, such as birds and pigs, are reservoirs for influenza virus, viral surveillance is an important way to monitor for dangerous viral reassortment [31]. Approaches to monitor influenza virus in animals have been developed, such as the One Health program, in order to better understand how virus can spread to humans thereby affecting public health [45,46]. Containing the spread of influenza virus through vaccination efforts, especially in low- and middle-income countries, is a key way to help protect public health of the population, but in particular children and pregnant women.

2.2. The Host Immune Response to Influenza Virus Infection

The host response to influenza virus can limit viral infection within the lung, thereby protecting host health. However, failure to appropriately regulate this response can lead to damage of lung alveoli due to excessive inflammation or cytolysis of lung cells due to viral infection. The result is impaired gas exchange, which is a major morbidity associated with influenza virus infection. Figure 2 summarizes factors that are responsible for protecting the host against infection with influenza virus. These include epithelial cells, cells of the innate immune system, the adaptive immune system, cytokines, chemokines, antibodies, and surfactant proteins.

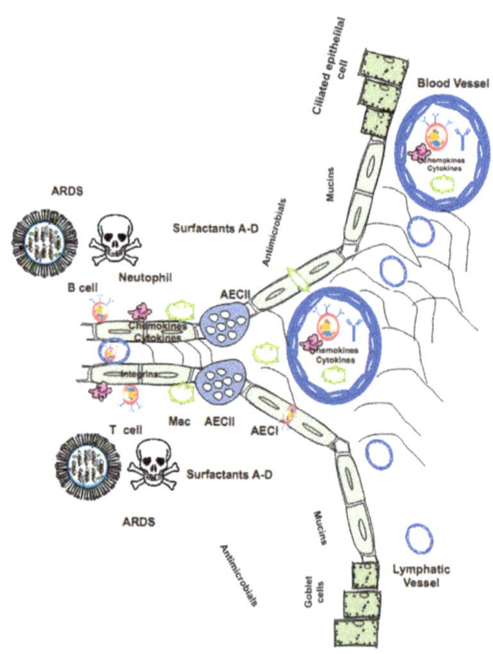

Figure 2. Factors involved in defense against influenza virus infection, which can also contribute to lung damage and ARDS. Cells of the immune system (e.g., macrophages; T cells; B cells; neutrophils) and lung (e.g., alveolar epithelial type I AECI; alveolar epithelial type II AECII; ciliated cells; goblet cells) in addition to factors such as surfactants, mucins, and antimicrobial proteins, interact to protect the host from influenza virus. Direct cytopathic effects of influenza virus infection and excessive inflammation lead to damage of alveoli, which can compromise respiratory function.

Influenza virus first targets epithelial cells of the proximal respiratory system through HA binding to α2,6-linked sialylated proteins (reviewed in [21,47,48]). Binding and/or internalization of the virus to cells of the epithelium leads to intracellular signaling that alters ion transport, contributing to symptoms of infection [49]. Antiviral responses are initiated, including the release of antimicrobial peptides such as surfactants, mucins, LL-37 and β-defensins, which decrease viral binding to epithelial cells and promote recruitment of innate immune cells such as neutrophils [47,50,51]. Surfactant proteins are capable of binding to virus, which helps to limit infectivity and disease severity [52–55]. Upon infection, respiratory epithelial cells sense virus through Toll-Like Receptors (TLRs), retinoic acid-inducible gene I (RIG-I), NOD-like receptors (NLRs), and melanoma differentiation-associated 5 (MDA-5), leading to the expression of type-I and type-III interferons (IFN), interleukin-6 (IL-6), IL-1β, IL-18, and other pro-inflammatory cytokines and chemokines [56–59]. Some of these cytokines and chemokines cause immune cells to extravasate from blood vessels into the site of infection in order to combat the pathogen.

Following influenza virus infection, a classic antiviral response occurs. Among the first cells to become activated are macrophages and dendritic cells, which are critical for the initiation of the antiviral response and instruction of developing adaptive immunity [60,61]. Many subtypes of dendritic cells are present, some of which are capable of presenting antigen to both CD4+ and CD8+ T cells [62]. Delivery of antigen to draining lymph nodes via lymphatic vessels is an important step in activating T cells [63,64]. Dendritic cells are targeted by influenza virus infection, which can impair the development of the adaptive immune response [65]. Once activated, CD8+ T cells will kill virally infected cells, with an important role for lung tissue-resident memory CD8 T cells (T_{rm}) in generating rapid antiviral responses upon host reinfection [66,67]. CD4+ T cells also contribute a multiplicity of functions to anti-influenza immunity, including promoting CD8+ T cell function (activation, expansion, positioning, and memory formation), the innate immune response, help for the B cell response, and independent cytotoxicity [68,69]. T follicular helper cells, a specialized CD4+ T cell subset, provide cognate help for both the extrafollicular and germinal center B cell responses [70–72]. Antibody secreting cells (ASCs) then home to tissues such as the bone marrow, where they receive survival factors that allow for long term survival, imparting immunity via the high levels of antibody they secrete [73]. An alternate fate upon B cell activation is to become a memory B cell [74–76]. In the case of influenza virus, T cells and B cells recognize viral components including both the surface proteins (HA and NA) and the internal virion proteins (NP, NS1, and M1); see Figure 1 [69]. Detectable levels of class-switched antibody are found approximately two weeks following influenza infection [77,78]. These high-affinity antibodies can then act to interfere with viral binding, viral replication, or target infected cells for killing via mechanisms such as antibody-dependent cellular cytotoxicity (ADCC) [79,80]. Following successful viral clearance, cells of the immune system become relatively quiescent once again, leaving a pool of memory T and B cells in addition to protective antibodies that will lead to resistance against future influenza infection [81–83].

2.3. Acute Respiratory Distress Syndrome and Lung Damage as a Consequence of Overwhelming Viral Infection

If the host is unable to control viral infection, respiratory morbidities can occur. One such morbidity is the development of ARDS, which is characterized by pulmonary edema, hypoxemia and a high mortality rate [15]. Infectious agents are just one of many potential causes of ARDS, with both direct influenza virus infection and the anti-influenza immune response contributing to damage of the respiratory tract and ARDS development [17]. In pregnant women and young children, ARDS is a rare but major health concern following infection with influenza virus [14,15], with one study demonstrating influenza virus infection to be a factor contributing to a higher risk of death in pregnant women with ARDS [16].

There are many examples of how viral and host factors promote the generation of inflammation in the lung, which is a hallmark of severe influenza-like illness. One report demonstrated that the glycosylation state of HA controlled levels of proinflammatory cytokines produced by human lung epithelial cells [84]. Also, influenza viruses that were selected to infect human epithelial cells became more pathogenic by adaptation to the host, thereby increasing illness severity [85]. It is well known that viral factors, such as NS1, downregulate the production of interferons, thereby decreasing the host antiviral response [86]. While gas-exchanging alveolar type I epithelial cells can generate an antiviral response, high viral loads can overwhelm this and lead to compromised lung function [87,88]. In some severe cases of influenza virus infection, virus and inflammation were noted in the lower respiratory tract or alveoli, with resulting cellular damage (Figure 2) [8,9,11–13,89]. Additionally, neutrophils have been implicated in the pathophysiology of alveolar damage following infection with influenza virus [90]. IL-8/CXCL8 and GM-CSF, which are neutrophil chemotactic agents, are produced by primary human alveolar epithelial cells and may exacerbate the inflammatory process in alveoli, thus increasing the likelihood of developing ARDS [91]. TLR3-expressing CD8+ T cells were also found in areas with diffuse alveolar damage in a group of patients who died of severe influenza virus infection [92]. These findings suggest several potential interventions aimed at limiting alveolar damage through control of excessive inflammation that may help to reduce the risk of developing ARDS.

Controlling viral load and the resulting pulmonary inflammation through vaccination pre-exposure and use of therapeutics post-exposure could help to decrease lung damage and the resulting respiratory morbidity. In support of this, ferrets and mice that were administered human immunoglobulins had a reduction in viral load as compared to control animals that did not [93], with decreased weight loss, decreased mortality, and protection against challenge with H1N1 virus [94]. Similarly, data suggested that a high viral load and a strong cytokine response contributed to mortality in humans infected with H5N1 influenza virus [95]. One recent study demonstrated that children between one and five years of age had higher viral loads compared to older children, perhaps contributing to their increased susceptibility to influenza virus infection [96]. While these results seem to indicate that decreasing viral load would lead to improved outcomes, a recent phase III clinical trial demonstrated that transfer of sera containing high levels of anti-influenza antibody into patients with severe influenza virus infection did not lead to detectable benefit [97]. However, administration of some anti-influenza virus antibodies have been shown to reduce viral loads in humans [98]. Additional work will be required to determine whether antibody treatments will help to reduce influenza virus load and decrease disease severity.

Following clearance of the infection, lung repair occurs via a complicated interplay between the immune system and epithelium [99–102]. Ineffective repair of alveoli following infection has also been implicated in the development of ARDS [15]. Thus, the interaction between the lung and immune system is the key not only to anti-viral responses, but also for post-infection healing. Unfortunately, basic information regarding the interaction of cells found in human lung is lacking in the field of lung biology. Work from our group is attempting to unravel the crosstalk between epithelial cells and immune cells in the developing human lung [103–105]. Such information and the experimental models that have been developed will be useful in future studies aimed at identifying treatments to prevent ARDS or encourage lung healing.

2.4. *The Pediatric Immune System is Different than that of the Adult*

Differences in immunologic function between young children and adults likely contribute to the increased susceptibility of children to influenza virus infection. The immune system of young children is characterized by a higher frequency of naïve antigen-specific cells [106,107]. This population also tends to have an overall higher number of circulating T and B cells that decreases drastically by six years of age [108]. An increase in regulatory T and B cells is also noted in neonatal blood samples [109–113]. Despite the increase in the frequency of regulatory T cells in neonatal blood, the functional ability of neonatal regulatory T cells to suppress dendritic cell function and to blunt the immune reaction is diminished relative to adults [114]. Additionally, cells of the pediatric innate immune system, including monocytes and dendritic cells, tend to be less stimulatory (e.g., increased IL-10 and decreased IL-12, IL-1β, and IFN-α) when challenged with TLR agonists versus adults [115,116]. This leads to decreased antigen presentation and T cell co-stimulation capacity. Neutrophil function, such as the generation of reactive oxygen species and neutrophil extracellular traps, is also reduced in neonates relative to adults [117]. These characteristics of the pediatric immune system may contribute to the increased morbidity of children upon infection with influenza virus and the need for children to receive two doses of IIV vaccine. Efforts are underway to produce adjuvanted vaccines to help prime a more robust immune response (see below).

2.5. Alterations in the Immune System and Epithelial Barrier of the Preterm Infants Versus Term Infants

Infants born preterm (<37 weeks gestational age) face a variety of long-term health morbidities. While the threat of developing sepsis is a major concern for all neonates, the concern is even greater for preterm infants [118]. Hyporesponsiveness of monocytes to stimulation and decreased TNF-α responsiveness in non-classical monocyte activation have also been noted in preterm infants, especially prior to 30 weeks gestational age, and may contribute to sepsis risk [119,120]. Infections that are known to cause sepsis compromise the epithelial barrier of the lung and gut, leading to decreased expression of several TLRs and easier access of infectious agents to the underlying tissue, facilitating development of infections [121]. Furthermore, antimicrobial factors such as LL-37 are lower in preterm cord blood versus full term controls, which could be reflective of impaired antimicrobial potential in preterm infants [118]. Work involving our group has shown that preterm B cell responses to the influenza virus vaccine actually generate a higher peak level of influenza virus-specific antibody versus term controls, with levels equalizing by 9 months after vaccination [122,123]. We have also reported that CD4+ T cells found in cord blood from preterm infants who are exposed to inflammatory stimuli *in utero* are more likely to produce proinflammatory cytokines, including IL-6 [124]. These studies are just some examples of studies found in the literature that illustrate how the immune system of preterm infants is different from term infants.

An additional complication of preterm birth is abnormal lung development. Bronchopulmonary dysplasia (BPD) is a chronic lung disease characterized by inflammation and arrest of alveolar development that affects 30–60% of infants born preterm [125–128]. Studies suggest that part of the oxidative and mechanical damage is the result of respiratory ventilation [129]. One paper suggests that BPD could predispose infants to developing pediatric ARDS [130]. Of note, a recent study from Australia reports that children with BPD have an increased incidence ratio of being hospitalized due to influenza-related illness (9.0 ratio for children 0–10 years old and 41.6 for the 0–2 year old age range) [131]. Additionally, the length of hospital stay for children with BPD was 7 days longer than for children with cystic fibrosis or other chronic lung diseases. Thus, children who are born preterm and develop BPD are at particularly high risk for developing severe influenza illness.

2.6. Vaccination Strategies for Protecting Public Health Vary Across the World

Influenza vaccine is administered primarily through intramuscular injection (inactivated subunit or split virus) or spraying intranasally (live attenuated influenza virus). Typically, vaccines contain three or four strains of virus (trivalent or quadrivalent respectively), including two influenza A strains (H1 and H3), and one or two influenza B strains [18]. A vaccine is considered to confer protection if post-vaccination serum can inhibit influenza mediated hemagglutination *in vitro* at a one to forty or greater dilution, although higher HAI titers may be necessary to achieve similar levels of protection in children [132]. Ultimately, the goal of vaccination is to reduce the rate or severity of infection. This can occur by antibodies that interfere with viral binding to target cells (e.g., neutralization) or by inhibiting the "life-cycle" of the virus (e.g., by preventing viral release from infected cells). Antibody-dependent cellular cytotoxicity is another mechanism for killing virally infected cells, which illustrates another important function that antibodies have [79,80]. In addition to antibodies, a recall response of memory CD4+ and CD8+ T cells allows for protection against influenza virus infection [133,134].

Recommendations for vaccinating children and pregnant women vary around the world. In the United States and Canada, it is recommended that all infants and children greater than 6 months of age be vaccinated using IIV, with administration of LAIV only in children greater than 24 months of age due to a reported increased risk of post-LAIV wheezing in infants [135,136]. As children may be relatively immunologically naïve to influenza virus, those younger than 9 years of age initially receive a priming dose of vaccine followed by a booster vaccine dose at least 28 days later [136]. In addition, IIV administration is recommended for women who are or will be pregnant during the influenza season [137]. Although the immune system of preterm infants is relatively immature, immunizations are typically given at the same chronological age as in term infants [138,139].

Despite broad recommendations for influenza vaccination in some countries, the World Health Organization (WHO) reported that in 2014, only 59% of member countries had vaccination programs in place [140]. Given that 41% of countries did not have vaccination programs in place, the chances for the spread of influenza virus is quite high. Viral reassortment can occur in birds and pigs, which has contributed to the generation of novel pandemic influenza virus strains [141,142]. By decreasing the number of people infected via routine vaccination, there is likely to be a decreased chance for coinfection of birds and pigs with human and animal influenza virus, thus decreasing the chances of reassortment mutants that could become pandemic. Furthermore, if a country has a vaccination plan in place, rapid administration of vaccines in the case of a pandemic would be possible [143]. Amongst members of the European Union there is great variability in influenza vaccination requirements [144], with a set goal of a 75% vaccination rate in the high-risk population, but actual vaccination rates that are much lower [145]. Countries with low- to middle-income typically have low vaccination rates against influenza virus [146]. Individuals living in these countries face a higher disease burden than individuals from higher-income countries [147]. In countries that do have influenza immunization programs, fewer than half recommended vaccinating pregnant women and less than a third recommended routine vaccination of children [140]. Regardless of recommendations, influenza vaccination rates remain suboptimal and demonstrate substantial variability by age, location, and season. In the US, the goal is to vaccinate 70% of the population, but only around 40% of adults were vaccinated each year between 2010 and 2016 [148]. Similarly, one Canadian study reported that, on average, only 29% of respondents had been vaccinated in a given year between the 2006 and 2013 influenza seasons [149]. Furthermore, one study shows that less than half of pregnant women in the United States were vaccinated over three consecutive influenza seasons [150]. Highlighting the importance of vaccination, a study of 358 laboratory-confirmed influenza-associated pediatric deaths between 2010 and 2014 reported that vaccination coverage was low (26% overall and only 31% of high-risk patients) [151]. Unfortunately, children delivered prematurely have higher health risks associated with birth during the influenza season and a higher risk for being admitted into the hospital than full-term children [131,152,153], yet a recent study demonstrated that late preterm infants were less likely than term infants to be appropriately vaccinated against influenza by 36 months of age [154]. These data highlight the importance of efforts to increase overall influenza vaccination rates among both pregnant women and children.

2.7. A Key Role for Maternal Vaccination in Protecting Young Infants Against Influenza

The World Health Organization has identified pregnant women as a priority group for receiving influenza vaccines [155]. Pregnant women are particularly susceptible to developing severe influenza illness and have an increased risk for hospital admission, although the mortality rate compared to non-pregnant women may not be increased [156–158]. Overall, it has been established that vaccinating pregnant women is safe, with no evidence demonstrating a link between immunization during pregnancy and adverse outcomes in offspring [159–165]. Infants born to mothers who were vaccinated against influenza may have a decreased rate of preterm birth, low birthweight, and stillborn birth, although establishing a causal relationship is challenging [160,166–171]. Infants born to mothers who were severely sick with H1N1 influenza infection have been shown to have an increased risk of preterm birth, decreased 5 min Apgar scores, and an increased risk of death [172–174]. These data support vaccinating pregnant women to help improve the health of both the mother and child.

Benefits to the fetus from maternal vaccination against influenza virus include transfer of maternal antibody through the placenta. This is critical for protecting neonates and infants less than six months of age against severe influenza virus infection requiring hospitalization, as these infants are too young to be vaccinated [7,153]. Studies have demonstrated that some isotypes of anti-influenza antibodies cross the placenta and result in higher HAI titers in cord blood compared to placebo recipients [175], with protective antibody transported to the fetus in as little as two weeks post-vaccination [176]. Transplacental transport of antibodies requires expression of the neonatal Fc receptor expression, with IgG, especially IgG1, being effectively transported into fetal circulation [177,178]. Importantly, decreased rates of acute lower respiratory tract infection, influenza virus infections, and influenza-related hospitalizations were reported in infants following maternal vaccination during pregnancy [179–181]. While vaccination at any point during pregnancy is recommended, children born to mothers who were vaccinated more than 4 weeks prior to delivery and during the second or third trimester had higher antibody titers against the A(H1N1) virus when compared to those who were born to mother that were vaccinated in the first trimester [182].

Another benefit of maternal vaccination is transfer of protective antibodies to children via breastmilk. This is particularly important for antibodies of the IgA isotype, which are not passed through the placenta [183,184]. Vaccination of breastfeeding mothers with IIV resulted in significantly higher HAI titers in serum and IgG and IgA levels in breast milk as compared to vaccination using LAIV, suggesting that IIV may be the preferred vaccine for use in breastfeeding mothers [185]. Given that preterm infants have lower levels of IgG antibody transferred via the placenta, antibody transferred via breastmilk may be of particular importance to decrease influenza virus infection in the preterm population [186,187]. Of note, milk from mothers who delivered prematurely had lower levels of total IgG and IgM but not IgA as compared to mothers who delivered at term [186]. Increasingly, preterm infants are being fed human donor breast milk, which contains significantly lower concentrations of IgM than non-donor milk, possibly due to the pasteurization process. However, administration of any antibody through donor breast milk would logically be better than receiving none at all. Several recent studies have focused on antibody transfer to the stomachs of preterm infants. Antibodies have been shown to be more stable in the gastric contents of preterm infants versus term infants [186], with influenza-virus-specific IgA antibodies being more stable in gastric contents of preterm infants regardless of whether milk is from the birth mother or from a donor [188].

In addition to antibodies, other bioactive molecules are found in breast milk, including cytokines and human milk oligosaccharides (HMOs) that have been shown to inhibit viral entry into cells [189–191]. Furthermore, cells are transferred to the infant via breastfeeding, which could play a role in shaping the neonatal and infant microbiome [192]. Of note, studies have shown that HMOs fed to mice could be transferred into circulation, had no observable adverse health effects and improved the immune response to influenza virus infection [193–195]. In adult humans, HMO ingestion was well tolerated in a two-week oral administration regiment [196]. Thus, breast milk offers a variety of factors in addition to antibodies that help to protect offspring against influenza virus infection. Taken together, the above factors contribute to improved maternal and fetal health afforded by vaccinating pregnant women against influenza virus.

2.8. Adjuvanted Influenza Vaccines

Due to the overall poor immunogenicity of influenza vaccines, the use of immune adjuvants is being evaluated as a strategy to increase vaccine immunogenicity. Several different adjuvants have been tested in pre-clinical models and clinical trials. These include aluminum salt (alum) based approaches, oil and water emulsions (MF59, AS03, and AF03), innate immune cell receptor agonists (e.g., TLRs and the inflammasome), and virosomes (lipid bilayer droplets) [197]. These adjuvants utilize multiple mechanisms to increase the immunogencity of influenza vaccines, including activating antigen-presenting cells (APC), increasing antigen uptake by APCs, and recruiting immune cells to the site of vaccine administration [115,116]. Several clinical trials have been conducted in adults and children indicate that the use of AS03 as an adjuvant in influenza vaccines and increases influenza-virus-specific antibody response [198–204]. Clinical trials using MF-59 as an adjuvant also demonstrated safety and improved rates of seroconversion in preterm and term infants [205], with more robust antiviral responses, activation of dendritic cells and an increased CD4+ T cell cytokine response following administration of an adjuvanted vaccine [206–208]. MF59 has been approved for use in the United States since 2015 and its use appears to not increase adverse reaction rates [209]. Additionally, several studies have examined the use of adjuvanted H1N1 vaccines in pregnant women and have found that these vaccines were well tolerated [210–212]. Although the increased immunogenicity is strongly advantageous, there is concern for potential adverse effects, including a possible association between an AS03-adjuvanted monovalent pandemic influenza vaccine and narcolepsy [213–215]. A recent paper reports a higher rate of adverse events reported when the vaccine was administered to patients outside of the recommended age groups [209]. Reports of adverse events have likely slowed the uptake of adjuvanted influenza vaccines in vulnerable populations such as pregnant women and children.

2.9. Increasing Vaccination Rates for Influenza

Vaccines must be considered effective at preventing disease and be regarded as safe [216]. However, adverse reactions do occur and can include fever, febrile seizures, hypersensitivity reactions, and possibly a small increase in the risk of developing Guillain-Barre syndrome [217,218]. Several organizations, including the WHO, the European Union, the US government, and independent agencies have mechanisms in place for tracking vaccine safety [216,219,220]. In the United States, the Vaccine Adverse Event Reporting System passively documents post-licensure adverse events following vaccination administration [221]. In Europe, the European Medicines Agency is responsible for tracking adverse events through the EudraVigilance Program [222]. Such data are valuable, however, a deeper understanding of additional factors that could have contributed to an adverse event is necessary in order to minimize the risk of drawing an invalid conclusion associating an adverse event with vaccination. Vaccinating pregnant women is generally regarded as safe [159–165]. However, some hesitancy over influenza vaccination is present in this population and the benefit of protecting both the mother and the fetus needs to be conveyed as outweighing the risk of potential side effects [223–226].

Perhaps not surprisingly, preemptive action is more cost-effective than is reactive action for a pandemic, although modeling suggests mortality rates don't differ between preemptive and reactive responses [227]. It is estimated that in order to prevent one case of influenza, five individuals would need to be vaccinated with IIV and seven with LAIV [9]. The importance of having a well-vaccinated population is at the heart of "herd immunity", where disruption of influenza transmission will lead to a less severe outbreak [228]. Herd immunity in children has been modeled, with studies suggesting that a substantial decrease in influenza infection of non-vaccinated individuals is predicted to occur, but only at very high rates of vaccine coverage (i.e., 90%) [229]. Development of more effective vaccines and increasing the rate of vaccination will bring the population closer protection from influenza virus infection [230]. Given the strong benefit of vaccinating pregnant women and children and the low vaccination rates across the globe, finding effective ways to increase these vaccination rates is critical for improving public health.

One major reason for vaccine refusal relates to the public's lack of trust in the pharmaceutical industry, medical providers, and efficacy of the influenza vaccine [231]. Some work suggests that mandatory vaccination increases negative feelings and anger regarding vaccines as compared to voluntary vaccination programs [232]. Such feelings could also lead to a stronger aversion to all vaccines [233]. A particularly influential factor that has led to mistrust of vaccine safety has been rooted in the belief that vaccination is associated with children being placed on the autism spectrum [234]. Initial studies making this claim have since been retracted, and subsequent work has not supported this association [235,236]. Despite this fact, recent studies continue to demonstrate that younger siblings of children on the autism spectrum disorder are not fully vaccinated [237,238]. This work illustrates the long-lasting harm caused by the initial reports linking vaccine administration to autism. Furthermore, it shows that the general public is still very much concerned with the perceived risk of administering vaccines to children despite a lack of supporting evidence. Rebuilding confidence in vaccines will require communication and trust between clinical providers and patients [239,240].

Increasing vaccination rates in low- and middle-income countries will require additional resources to help offset the cost of development and implementation of vaccination programs [225]. It has been estimated that global supply of influenza virus vaccine is sufficient to cover pregnant women in low- and middle-income countries [241]. A recent phase 4 clinical trial demonstrated that developing a vaccination program for pregnant women is possible in Mali, a low-income country [242,243]. While the Mali study was not sufficiently powered to detect a decrease in ILI, maternal vaccination studies conducted in Bangladesh and South Africa have demonstrated at least partial protection against developing confirmed influenza virus infection [175,180]. Unfortunately, due to the lack of sufficient data for the benefit of developing such programs, the Global Alliance for Vaccines and Immunization has deprioritized maternal vaccinations against seasonal flu [244–247] while the Global Influenza Initiative recommends that all pregnant women be vaccinated in their third trimester [225]. This lack of harmonization between guidelines highlights the additional work needed to develop evidence-based recommendations for programs in resource poor countries [248]. Regardless, by increasing vaccine uptake in low- and middle-income countries, not only will human health be improved, but the chances of virus spreading to an expanded geographical region are lessened. This will help to contain virus and likely decrease the chance of establishing a pandemic.

3. Discussion

Influenza infection is a serious health concern, especially for pregnant women and young children. While influenza vaccines are generally considered to be safe, vaccine uptake remains suboptimal. Vaccination of pregnant women provides protection against influenza infection in both the expectant mother as well as the infant due to transplacental transfer of influenza-virus-specific IgG antibody. Additionally, breastfeeding provides antibodies (in particular IgA that is not passed through the placenta) and immunomodulatory factors to prevent and/or combat influenza infection. These factors are particularly important in the first six months of life, as active vaccination is not recommended for this age group. For infants born prematurely, complicating factors such as chronic lung disease increases the risk for developing severe illness after influenza infection. Despite having an immature immune system, vaccination is recommended on schedule in this population and has been shown to be protective. Furthermore, vaccination of these populations will help to guard against the development of ARDS, which is a major health concern following infection with influenza virus.

Influenza viruses undergoes continuous antigenic drift, which leads to lower than ideal vaccine efficacy in some seasons. Furthermore, antigenic shifts resulting in pandemic outbreaks is not uncommon, with 4 pandemics in the 21st century. Next generation influenza vaccines targeted against highly conserved regions of the influenza virus are being developed that may provide more universal protection against even potentially pandemic influenza strains. However, even if conventional vaccine strains of influenza are not well matched to circulating strains, reduced viral shedding and a shorter duration and severity of illness are often observed in the vaccinated population. Such information must be effectively disseminated to members of the public in order to improve vaccine uptake, as misconceptions about influenza vaccine adverse effects and effectiveness remains major obstacles to improving worldwide influenza vaccination rates.

4. Conclusions

Influenza vaccination reduces the risk of influenza infection, severe disease, morbidity, and death. Despite this, the rate of influenza vaccination remains well below targets set forth by international and national health officials. While improving vaccination rates, especially among pregnant women and young children, is a high priority, additional large-scale studies would be of benefit to generate evidence-based recommendations in support of existing programs and to increase public confidence in current recommendations. In addition, development of more immunogenic and universally protective influenza vaccines will increase the breadth of protection provided while decreasing the frequency of vaccination required. Such efforts will be critical to increasing uptake of influenza vaccines and will improve the health of at-risk populations, such as children and pregnant women.

Author Contributions: Conceptualization, R.S.M. and J.L.N.; writing—original draft preparation, R.S.M.; writing—review and editing, R.S.M. and J.L.N.; funding acquisition, R.S.M.

Funding: This research was funded by The University of Rochester Department of Pediatrics.

Acknowledgments: The author thanks Gloria Pryhuber and Carl D'Angio in their support in his writing of the review article. Figures were generated using EDraw Max software.

Conflicts of Interest: The authors declare no conflict of interest.

Abbreviations

ARDS	Acute respiratory distress syndrome
IIV	Inactivated Influenza Vaccine
LAIV	Live Attenuated Influenza Virus
HA	hemagglutinin
NA	neuraminidase
NP	nucleoprotein
M1	matrix protein 1
M2/BM2	matrix protein 2
NS1	nonstructural protein 1
NS2	nonstructural protein 2
TLRs	Toll-Like Receptors
RIG-I	retinoic acid-inducible gene I
NLRs	NOD-like receptors
MDA-5	melanoma differentiation-associated 5
IFN	interferons
IL	interleukin
ASCs	antibody secreting cells
ADCC	antibody-dependent cellular cytotoxicity
WHO	World Health Organization
AECI	alveolar epithelial type I
AECII	alveolar epithelial type II

References

1. Doyon-Plourde, P.; Fakih, I.; Tadount, F.; Fortin, E.; Quach, C. Impact of influenza vaccination on healthcare utilization—A systematic review. *Vaccine* **2019**, *37*, 3179–3189. [CrossRef] [PubMed]
2. United_States_Food_and_Drug_Administration. Pediatric Medical Devices. Available online: https://www.fda.gov/medical-devices/products-and-medical-procedures/pediatric-medical-devices (accessed on 25 November 2019).
3. Nair, H.; Brooks, W.A.; Katz, M.; Roca, A.; Berkley, J.A.; Madhi, S.A.; Simmerman, J.M.; Gordon, A.; Sato, M.; Howie, S.; et al. Global burden of respiratory infections due to seasonal influenza in young children: A systematic review and meta-analysis. *Lancet* **2011**, *378*, 1917–1930. [CrossRef]
4. Nunes, M.C.; Madhi, S.A. Influenza vaccination during pregnancy for prevention of influenza confirmed illness in the infants: A systematic review and meta-analysis. *Hum. Vaccines Immunother.* **2018**, *14*, 758–766. [CrossRef] [PubMed]
5. Lafond, K.E.; Nair, H.; Rasooly, M.H.; Valente, F.; Booy, R.; Rahman, M.; Kitsutani, P.; Yu, H.; Guzman, G.; Coulibaly, D.; et al. Global Role and Burden of Influenza in Pediatric Respiratory Hospitalizations, 1982–2012: A Systematic Analysis. *PLoS Med.* **2016**, *13*, e1001977. [CrossRef] [PubMed]

6. Committee On Infectious Diseases. Recommendations for Prevention and Control of Influenza in Children, 2018–2019. *Pediatrics* **2018**, *142*, e20182367. [CrossRef]
7. Nunes, M.C.; Madhi, S.A. Prevention of influenza-related illness in young infants by maternal vaccination during pregnancy. *F1000Research* **2018**, *7*, 122. [CrossRef]
8. Short, K.R.; Kroeze, E.; Fouchier, R.A.M.; Kuiken, T. Pathogenesis of influenza-induced acute respiratory distress syndrome. *Lancet Infect. Dis.* **2014**, *14*, 57–69. [CrossRef]
9. Simonsen, L.; Clarke, M.J.; Schonberger, L.B.; Arden, N.H.; Cox, N.J.; Fukuda, K. Pandemic versus epidemic influenza mortality: A pattern of changing age distribution. *J. Infect. Dis.* **1998**, *178*, 53–60. [CrossRef]
10. Short, K.R.; Kasper, J.; van der Aa, S.; Andeweg, A.C.; Zaaraoui-Boutahar, F.; Goeijenbier, M.; Richard, M.; Herold, S.; Becker, C.; Scott, D.P.; et al. Influenza virus damages the alveolar barrier by disrupting epithelial cell tight junctions. *Eur. Respir. J.* **2016**, *47*, 954–966. [CrossRef]
11. Wonderlich, E.R.; Swan, Z.D.; Bissel, S.J.; Hartman, A.L.; Carney, J.P.; O'Malley, K.J.; Obadan, A.O.; Santos, J.; Walker, R.; Sturgeon, T.J.; et al. Widespread Virus Replication in Alveoli Drives Acute Respiratory Distress Syndrome in Aerosolized H5N1 Influenza Infection of Macaques. *J. Immunol.* **2017**, *198*, 1616–1626. [CrossRef]
12. Kennedy, E.D.; Roy, M.; Norris, J.; Fry, A.M.; Kanzaria, M.; Blau, D.M.; Shieh, W.J.; Zaki, S.R.; Waller, K.; Kamimoto, L.; et al. Lower respiratory tract hemorrhage associated with 2009 pandemic influenza A (H1N1) virus infection. *Influenza Other Respir. Viruses* **2013**, *7*, 761–765. [CrossRef] [PubMed]
13. Shieh, W.J.; Blau, D.M.; Denison, A.M.; Deleon-Carnes, M.; Adem, P.; Bhatnagar, J.; Sumner, J.; Liu, L.; Patel, M.; Batten, B.; et al. 2009 pandemic influenza A (H1N1): Pathology and pathogenesis of 100 fatal cases in the United States. *Am. J. Pathol.* **2010**, *177*, 166–175. [CrossRef] [PubMed]
14. Nye, S.; Whitley, R.J.; Kong, M. Viral Infection in the Development and Progression of Pediatric Acute Respiratory Distress Syndrome. *Front. Pediatr.* **2016**, *4*, 128. [CrossRef] [PubMed]
15. Matthay, M.A.; Zemans, R.L.; Zimmerman, G.A.; Arabi, Y.M.; Beitler, J.R.; Mercat, A.; Herridge, M.; Randolph, A.G.; Calfee, C.S. Acute respiratory distress syndrome. *Nat. Rev. Dis Primers* **2019**, *5*, 18. [CrossRef]
16. Rush, B.; Martinka, P.; Kilb, B.; McDermid, R.C.; Boyd, J.H.; Celi, L.A. Acute Respiratory Distress Syndrome in Pregnant Women. *Obstet. Gynecol.* **2017**, *129*, 530–535. [CrossRef]
17. Kalil, A.C.; Thomas, P.G. Influenza virus-related critical illness: Pathophysiology and epidemiology. *Crit. Care* **2019**, *23*, 258. [CrossRef]
18. Mameli, C.; Cocchi, I.; Fumagalli, M.; Zuccotti, G. Influenza Vaccination: Effectiveness, Indications, and Limits in the Pediatric Population. *Front. Pediatri.* **2019**, *7*, 317. [CrossRef]
19. De, A. Molecular evolution of hemagglutinin gene of Influenza A virus. *Front. Biosci.* **2018**, *10*, 101–118. [CrossRef]
20. Arai, Y.; Kawashita, N.; Hotta, K.; Hoang, P.V.M.; Nguyen, H.L.K.; Nguyen, T.C.; Vuong, C.D.; Le, T.T.; Le, M.T.Q.; Soda, K.; et al. Multiple polymerase gene mutations for human adaptation occurring in Asian H5N1 influenza virus clinical isolates. *Sci. Rep.* **2018**, *8*, 13066. [CrossRef]
21. de Graaf, M.; Fouchier, R.A. Role of receptor binding specificity in influenza A virus transmission and pathogenesis. *EMBO J.* **2014**, *33*, 823–841. [CrossRef]
22. Goka, E.A.; Vallely, P.J.; Mutton, K.J.; Klapper, P.E. Mutations associated with severity of the pandemic influenza A(H1N1)pdm09 in humans: A systematic review and meta-analysis of epidemiological evidence. *Arch. Virol.* **2014**, *159*, 3167–3183. [CrossRef] [PubMed]
23. Simon, B.; Pichon, M.; Valette, M.; Burfin, G.; Richard, M.; Lina, B.; Josset, L. Whole Genome Sequencing of A(H3N2) Influenza Viruses Reveals Variants Associated with Severity during the 2016–2017 Season. *Viruses* **2019**, *11*, 108. [CrossRef] [PubMed]
24. Arai, Y.; Kawashita, N.; Ibrahim, M.S.; Elgendy, E.M.; Daidoji, T.; Ono, T.; Takagi, T.; Nakaya, T.; Matsumoto, K.; Watanabe, Y. PB2 mutations arising during H9N2 influenza evolution in the Middle East confer enhanced replication and growth in mammals. *PLoS Pathog.* **2019**, *15*, e1007919. [CrossRef] [PubMed]
25. Seladi-Schulman, J.; Campbell, P.J.; Suppiah, S.; Steel, J.; Lowen, A.C. Filament-producing mutants of influenza A/Puerto Rico/8/1934 (H1N1) virus have higher neuraminidase activities than the spherical wild-type. *PLoS ONE* **2014**, *9*, e112462. [CrossRef]
26. Wang, Q.; Bhattacharya, S.; Mereness, J.A.; Anderson, C.; Lillis, J.A.; Misra, R.S.; Romas, S.; Huyck, H.; Howell, A.; Bandyopadhyay, G.; et al. A novel in vitro model of primary human pediatric lung epithelial cells. *Pediatr. Res.* **2019**. [CrossRef]

27. Thangavel, R.R.; Reed, A.; Norcross, E.W.; Dixon, S.N.; Marquart, M.E.; Stray, S.J. "Boom" and "Bust" cycles in virus growth suggest multiple selective forces in influenza a evolution. *Virol. J.* **2011**, *8*, 180. [CrossRef]
28. Zhou, F.; Trieu, M.C.; Davies, R.; Cox, R.J. Improving influenza vaccines: Challenges to effective implementation. *Curr. Opin. Immunol.* **2018**, *53*, 88–95. [CrossRef]
29. Valkenburg, S.A.; Leung, N.H.L.; Bull, M.B.; Yan, L.M.; Li, A.P.Y.; Poon, L.L.M.; Cowling, B.J. The Hurdles From Bench to Bedside in the Realization and Implementation of a Universal Influenza Vaccine. *Front. Immunol.* **2018**, *9*, 1479. [CrossRef]
30. Epstein, S.L. Universal Influenza Vaccines: Progress in Achieving Broad Cross-Protection In Vivo. *Am. J. Epidemiol.* **2018**, *187*, 2603–2614. [CrossRef]
31. Krammer, F.; Smith, G.J.D.; Fouchier, R.A.M.; Peiris, M.; Kedzierska, K.; Doherty, P.C.; Palese, P.; Shaw, M.L.; Treanor, J.; Webster, R.G.; et al. Influenza. *Nat. Rev. Dis. Primers* **2018**, *4*, 3. [CrossRef]
32. Sharma, L.; Rebaza, A.; Dela Cruz, C.S. When "B" becomes "A": The emerging threat of influenza B virus. *Eur. Respir. J.* **2019**, *54*, 1901326. [CrossRef] [PubMed]
33. Tan, J.; Asthagiri Arunkumar, G.; Krammer, F. Universal influenza virus vaccines and therapeutics: Where do we stand with influenza B virus? *Curr. Opin. Immunol.* **2018**, *53*, 45–50. [CrossRef] [PubMed]
34. Proff, R.; Gershman, K.; Lezotte, D.; Nyquist, A.C. Case-based surveillance of influenza hospitalizations during 2004–2008, Colorado, USA. *Emerg. Infect. Dis.* **2009**, *15*, 892–898. [CrossRef] [PubMed]
35. Chiu, S.S.; Chan, K.H.; Chen, H.; Young, B.W.; Lim, W.; Wong, W.H.; Lau, Y.L.; Peiris, J.S. Virologically confirmed population-based burden of hospitalization caused by influenza A and B among children in Hong Kong. *Clin. Infect. Dis.* **2009**, *49*, 1016–1021. [CrossRef]
36. Tran, D.; Vaudry, W.; Moore, D.; Bettinger, J.A.; Halperin, S.A.; Scheifele, D.W.; Jadvji, T.; Lee, L.; Mersereau, T.; Members of the Canadian Immunization Monitoring Program Active. Hospitalization for Influenza A Versus B. *Pediatrics* **2016**, *138*, e20154643. [CrossRef]
37. Schrauwen, E.J.; de Graaf, M.; Herfst, S.; Rimmelzwaan, G.F.; Osterhaus, A.D.; Fouchier, R.A. Determinants of virulence of influenza A virus. *Eur. J. Clin. Microbiol. Infect. Dis.* **2014**, *33*, 479–490. [CrossRef]
38. Wang, H.; Dou, D.; Ostbye, H.; Revol, R.; Daniels, R. Structural restrictions for influenza neuraminidase activity promote adaptation and diversification. *Nat. Microbiol.* **2019**. [CrossRef]
39. Morens, D.M.; Taubenberger, J.K.; Fauci, A.S. The persistent legacy of the 1918 influenza virus. *N. Engl. J. Med.* **2009**, *361*, 225–229. [CrossRef]
40. Lowen, A.C. Constraints, Drivers, and Implications of Influenza A Virus Reassortment. *Annu. Rev. Virol.* **2017**, *4*, 105–121. [CrossRef]
41. Potter, C.W.; Jennings, R. A definition for influenza pandemics based on historical records. *J. Infect.* **2011**, *63*, 252–259. [CrossRef]
42. Werner, J.; Schudrowitz, C.; Kohler, H. Antigenic variation of neuraminidase of human type A influenza (H3N2) viruses isolated in Berlin (West). *Zentralbl Bakteriol Orig A* **1975**, *233*, 440–446. [PubMed]
43. Kelly, H. The classical definition of a pandemic is not elusive. *Bull. World Health Organ.* **2011**, *89*, 540–541. [CrossRef] [PubMed]
44. Doshi, P. The elusive definition of pandemic influenza. *Bull. World Health Organ.* **2011**, *89*, 532–538. [CrossRef] [PubMed]
45. Zinsstag, J.; Schelling, E.; Waltner-Toews, D.; Tanner, M. From "one medicine" to "one health" and systemic approaches to health and well-being. *Prev. Vet. Med.* **2011**, *101*, 148–156. [CrossRef] [PubMed]
46. Short, K.R.; Richard, M.; Verhagen, J.H.; van Riel, D.; Schrauwen, E.J.; van den Brand, J.M.; Manz, B.; Bodewes, R.; Herfst, S. One health, multiple challenges: The inter-species transmission of influenza A virus. *One Health* **2015**, *1*, 1–13. [CrossRef] [PubMed]
47. Denney, L.; Ho, L.P. The role of respiratory epithelium in host defence against influenza virus infection. *Biomed. J.* **2018**, *41*, 218–233. [CrossRef]
48. Luo, M. Influenza virus entry. *Adv. Exp. Med. Biol.* **2012**, *726*, 201–221. [CrossRef]
49. Londino, J.D.; Lazrak, A.; Collawn, J.F.; Bebok, Z.; Harrod, K.S.; Matalon, S. Influenza virus infection alters ion channel function of airway and alveolar cells: Mechanisms and physiological sequelae. *Am. J. Physiol. Lung Cell. Mol. Physiol.* **2017**, *313*, L845–L858. [CrossRef]
50. Chen, Z.G.; Wang, Z.N.; Yan, Y.; Liu, J.; He, T.T.; Thong, K.T.; Ong, Y.K.; Chow, V.T.K.; Tan, K.S.; Wang, Y. Upregulation of cell-surface mucin MUC15 in human nasal epithelial cells upon influenza A virus infection. *BMC Infect. Dis.* **2019**, *19*, 622. [CrossRef]

51. Hsieh, I.N.; De Luna, X.; White, M.R.; Hartshorn, K.L. The Role and Molecular Mechanism of Action of Surfactant Protein D in Innate Host Defense Against Influenza A Virus. *Front. Immunol.* **2018**, *9*, 1368. [CrossRef]
52. Nikolaidis, N.M.; White, M.R.; Allen, K.; Tripathi, S.; Qi, L.; McDonald, B.; Taubenberger, J.; Seaton, B.A.; McCormack, F.X.; Crouch, E.C.; et al. Mutations flanking the carbohydrate binding site of surfactant protein D confer antiviral activity for pandemic influenza A viruses. *Am. J. Physiol. Lung Cell. Mol. Physiol.* **2014**, *306*, L1036–L1044. [CrossRef] [PubMed]
53. Herrera-Ramos, E.; Lopez-Rodriguez, M.; Ruiz-Hernandez, J.J.; Horcajada, J.P.; Borderias, L.; Lerma, E.; Blanquer, J.; Perez-Gonzalez, M.C.; Garcia-Laorden, M.I.; Florido, Y.; et al. Surfactant protein A genetic variants associate with severe respiratory insufficiency in pandemic influenza A virus infection. *Crit. Care* **2014**, *18*, R127. [CrossRef] [PubMed]
54. Benne, C.A.; Kraaijeveld, C.A.; van Strijp, J.A.; Brouwer, E.; Harmsen, M.; Verhoef, J.; van Golde, L.M.; van Iwaarden, J.F. Interactions of surfactant protein A with influenza A viruses: Binding and neutralization. *J. Infect. Dis.* **1995**, *171*, 335–341. [CrossRef] [PubMed]
55. To, K.K.W.; Zhou, J.; Song, Y.Q.; Hung, I.F.N.; Ip, W.C.T.; Cheng, Z.S.; Chan, A.S.F.; Kao, R.Y.T.; Wu, A.K.L.; Chau, S.; et al. Surfactant protein B gene polymorphism is associated with severe influenza. *Chest* **2014**, *145*, 1237–1243. [CrossRef] [PubMed]
56. Hamilton, J.R.; Sachs, D.; Lim, J.K.; Langlois, R.A.; Palese, P.; Heaton, N.S. Club cells surviving influenza A virus infection induce temporary nonspecific antiviral immunity. *Proc. Natl. Acad. Sci. USA* **2016**, *113*, 3861–3866. [CrossRef]
57. Chamberlain, N.; Korwin-Mihavics, B.R.; Nakada, E.M.; Bruno, S.R.; Heppner, D.E.; Chapman, D.G.; Hoffman, S.M.; van der Vliet, A.; Suratt, B.T.; Dienz, O.; et al. Lung epithelial protein disulfide isomerase A3 (PDIA3) plays an important role in influenza infection, inflammation, and airway mechanics. *Redox Biol.* **2019**, *22*, 101129. [CrossRef]
58. Sell, S.; McKinstry, K.K.; Strutt, T.M. Mouse Models Reveal Role of T-Cytotoxic and T-Reg Cells in Immune Response to Influenza: Implications for Vaccine Design. *Viruses* **2019**, *11*, 52. [CrossRef]
59. Kim, T.H.; Lee, H.K. Differential roles of lung dendritic cell subsets against respiratory virus infection. *Immune Netw.* **2014**, *14*, 128–137. [CrossRef]
60. He, W.; Chen, C.J.; Mullarkey, C.E.; Hamilton, J.R.; Wong, C.K.; Leon, P.E.; Uccellini, M.B.; Chromikova, V.; Henry, C.; Hoffman, K.W.; et al. Alveolar macrophages are critical for broadly-reactive antibody-mediated protection against influenza A virus in mice. *Nat. Commun.* **2017**, *8*, 846. [CrossRef]
61. Kumagai, Y.; Takeuchi, O.; Kato, H.; Kumar, H.; Matsui, K.; Morii, E.; Aozasa, K.; Kawai, T.; Akira, S. Alveolar macrophages are the primary interferon-alpha producer in pulmonary infection with RNA viruses. *Immunity* **2007**, *27*, 240–252. [CrossRef]
62. Vangeti, S.; Yu, M.; Smed-Sorensen, A. Respiratory Mononuclear Phagocytes in Human Influenza A Virus Infection: Their Role in Immune Protection and As Targets of the Virus. *Front. Immunol.* **2018**, *9*, 1521. [CrossRef] [PubMed]
63. Spitaels, J.; Roose, K.; Saelens, X. Influenza and Memory T Cells: How to Awake the Force. *Vaccines* **2016**, *4*, 33. [CrossRef] [PubMed]
64. Lambrecht, B.N.; Hammad, H. Lung dendritic cells in respiratory viral infection and asthma: From protection to immunopathology. *Annu. Rev. Immunol.* **2012**, *30*, 243–270. [CrossRef] [PubMed]
65. Smed-Sorensen, A.; Chalouni, C.; Chatterjee, B.; Cohn, L.; Blattmann, P.; Nakamura, N.; Delamarre, L.; Mellman, I. Influenza A virus infection of human primary dendritic cells impairs their ability to cross-present antigen to CD8 T cells. *PLoS Pathog.* **2012**, *8*, e1002572. [CrossRef] [PubMed]
66. Pizzolla, A.; Nguyen, T.H.O.; Smith, J.M.; Brooks, A.G.; Kedzierska, K.; Heath, W.R.; Reading, P.C.; Wakim, L.M. Resident memory CD8+ T cells in the upper respiratory tract prevent pulmonary influenza virus infection. *Sci. Immunol.* **2017**, *2*, eaam6970. [CrossRef] [PubMed]
67. Wu, T.; Hu, Y.; Lee, Y.T.; Bouchard, K.R.; Benechet, A.; Khanna, K.; Cauley, L.S. Lung-resident memory CD8 T cells (TRM) are indispensable for optimal cross-protection against pulmonary virus infection. *J. Leukoc. Biol.* **2014**, *95*, 215–224. [CrossRef]
68. Sant, A.J.; Richards, K.A.; Nayak, J. Distinct and complementary roles of CD4 T cells in protective immunity to influenza virus. *Curr. Opin. Immunol.* **2018**, *53*, 13–21. [CrossRef]

69. Sant, A.J.; DiPiazza, A.T.; Nayak, J.L.; Rattan, A.; Richards, K.A. CD4 T cells in protection from influenza virus: Viral antigen specificity and functional potential. *Immunol. Rev.* **2018**, *284*, 91–105. [CrossRef]
70. Biram, A.; Davidzohn, N.; Shulman, Z. T cell interactions with B cells during germinal center formation, a three-step model. *Immunol. Rev.* **2019**, *288*, 37–48. [CrossRef]
71. Takahashi, Y.; Onodera, T.; Adachi, Y.; Ato, M. Adaptive B Cell Responses to Influenza Virus Infection in the Lung. *Viral Immunol.* **2017**, *30*, 431–437. [CrossRef]
72. Shinnakasu, R.; Kurosaki, T. Regulation of memory B and plasma cell differentiation. *Curr. Opin. Immunol.* **2017**, *45*, 126–131. [CrossRef] [PubMed]
73. Nguyen, D.C.; Garimalla, S.; Xiao, H.; Kyu, S.; Albizua, I.; Galipeau, J.; Chiang, K.Y.; Waller, E.K.; Wu, R.; Gibson, G.; et al. Factors of the bone marrow microniche that support human plasma cell survival and immunoglobulin secretion. *Nat. Commun.* **2018**, *9*, 3698. [CrossRef] [PubMed]
74. Lavine, S.D.; Cockroft, K.; Hoh, B.; Bambakidis, N.; Khalessi, A.A.; Woo, H.; Riina, H.; Siddiqui, A.; Hirsch, J.A.; Chong, W.; et al. Erratum to: Training guidelines for endovascular stroke intervention: An international multi-society consensus document. *Neuroradiology* **2017**, *59*, 829. [CrossRef] [PubMed]
75. Ellebedy, A.H.; Jackson, K.J.; Kissick, H.T.; Nakaya, H.I.; Davis, C.W.; Roskin, K.M.; McElroy, A.K.; Oshansky, C.M.; Elbein, R.; Thomas, S.; et al. Defining antigen-specific plasmablast and memory B cell subsets in human blood after viral infection or vaccination. *Nat. Immunol.* **2016**, *17*, 1226–1234. [CrossRef]
76. Good-Jacobson, K.L. Strength in diversity: Phenotypic, functional, and molecular heterogeneity within the memory B cell repertoire. *Immunol. Rev.* **2018**, *284*, 67–78. [CrossRef]
77. Ellebedy, A.H. Immunizing the Immune: Can We Overcome Influenza's Most Formidable Challenge? *Vaccines* **2018**, *6*, 68. [CrossRef]
78. Lam, J.H.; Baumgarth, N. The Multifaceted B Cell Response to Influenza Virus. *J. Immunol.* **2019**, *202*, 351–359. [CrossRef]
79. Padilla-Quirarte, H.O.; Lopez-Guerrero, D.V.; Gutierrez-Xicotencatl, L.; Esquivel-Guadarrama, F. Protective Antibodies Against Influenza Proteins. *Front. Immunol.* **2019**, *10*, 1677. [CrossRef]
80. Vanderven, H.A.; Jegaskanda, S.; Wheatley, A.K.; Kent, S.J. Antibody-dependent cellular cytotoxicity and influenza virus. *Curr. Opin. Virol.* **2017**, *22*, 89–96. [CrossRef]
81. DiPiazza, A.; Richards, K.A.; Knowlden, Z.A.; Nayak, J.L.; Sant, A.J. The Role of CD4 T Cell Memory in Generating Protective Immunity to Novel and Potentially Pandemic Strains of Influenza. *Front. Immunol.* **2016**, *7*, 10. [CrossRef]
82. Dhume, K.; McKinstry, K.K. Early programming and late-acting checkpoints governing the development of CD4 T-cell memory. *Immunology* **2018**, *155*, 53–62. [CrossRef] [PubMed]
83. Auladell, M.; Jia, X.; Hensen, L.; Chua, B.; Fox, A.; Nguyen, T.H.O.; Doherty, P.C.; Kedzierska, K. Recalling the Future: Immunological Memory Toward Unpredictable Influenza Viruses. *Front. Immunol.* **2019**, *10*, 1400. [CrossRef] [PubMed]
84. Hrincius, E.R.; Liedmann, S.; Finkelstein, D.; Vogel, P.; Gansebom, S.; Samarasinghe, A.E.; You, D.; Cormier, S.A.; McCullers, J.A. Acute Lung Injury Results from Innate Sensing of Viruses by an ER Stress Pathway. *Cell Rep.* **2015**, *11*, 1591–1603. [CrossRef] [PubMed]
85. Huang, D.T.; Lu, C.Y.; Chi, Y.H.; Li, W.L.; Chang, L.Y.; Lai, M.J.; Chen, J.S.; Hsu, W.M.; Huang, L.M. Adaptation of influenza A (H7N9) virus in primary human airway epithelial cells. *Sci. Rep.* **2017**, *7*, 11300. [CrossRef]
86. Krug, R.M. Functions of the influenza A virus NS1 protein in antiviral defense. *Curr. Opin. Virol.* **2015**, *12*, 1–6. [CrossRef]
87. Sanders, C.J.; Vogel, P.; McClaren, J.L.; Bajracharya, R.; Doherty, P.C.; Thomas, P.G. Compromised respiratory function in lethal influenza infection is characterized by the depletion of type I alveolar epithelial cells beyond threshold levels. *Am. J. Physiol. Lung Cell. Mol. Physiol.* **2013**, *304*, L481–L488. [CrossRef]
88. Rosenberger, C.M.; Podyminogin, R.L.; Askovich, P.S.; Navarro, G.; Kaiser, S.M.; Sanders, C.J.; McClaren, J.L.; Tam, V.C.; Dash, P.; Noonan, J.G.; et al. Characterization of innate responses to influenza virus infection in a novel lung type I epithelial cell model. *J. Gen. Virol.* **2014**, *95*, 350–362. [CrossRef]
89. Gisslen, T.; Alvarez, M.; Wells, C.; Soo, M.T.; Lambers, D.S.; Knox, C.L.; Meinzen-Derr, J.K.; Chougnet, C.A.; Jobe, A.H.; Kallapur, S.G. Fetal inflammation associated with minimal acute morbidity in moderate/late preterm infants. Archives of disease in childhood. *Fetal Neonatal Ed.* **2016**, *101*, F513–F519. [CrossRef]

90. Tang, B.M.; Shojaei, M.; Teoh, S.; Meyers, A.; Ho, J.; Ball, T.B.; Keynan, Y.; Pisipati, A.; Kumar, A.; Eisen, D.P.; et al. Neutrophils-related host factors associated with severe disease and fatality in patients with influenza infection. *Nat. Commun.* **2019**, *10*, 3422. [CrossRef]
91. Ito, Y.; Correll, K.; Zemans, R.L.; Leslie, C.C.; Murphy, R.C.; Mason, R.J. Influenza induces IL-8 and GM-CSF secretion by human alveolar epithelial cells through HGF/c-Met and TGF-alpha/EGFR signaling. *Am. J. Physiol. Lung Cell. Mol. Physiol.* **2015**, *308*, L1178–L1188. [CrossRef]
92. Mauad, T.; Hajjar, L.A.; Callegari, G.D.; da Silva, L.F.; Schout, D.; Galas, F.R.; Alves, V.A.; Malheiros, D.M.; Auler, J.O., Jr.; Ferreira, A.F.; et al. Lung pathology in fatal novel human influenza A (H1N1) infection. *Am. J. Respir. Crit. Care Med.* **2010**, *181*, 72–79. [CrossRef] [PubMed]
93. Rockman, S.; Lowther, S.; Camuglia, S.; Vandenberg, K.; Taylor, S.; Fabri, L.; Miescher, S.; Pearse, M.; Middleton, D.; Kent, S.J.; et al. Intravenous Immunoglobulin Protects Against Severe Pandemic Influenza Infection. *EBioMedicine* **2017**, *19*, 119–127. [CrossRef] [PubMed]
94. Hohenadl, C.; Wodal, W.; Kerschbaum, A.; Fritz, R.; Howard, M.K.; Farcet, M.R.; Portsmouth, D.; McVey, J.K.; Baker, D.A.; Ehrlich, H.J.; et al. Hyperimmune intravenous immunoglobulin containing high titers of pandemic H1N1 hemagglutinin and neuraminidase antibodies provides dose-dependent protection against lethal virus challenge in SCID mice. *Virol. J.* **2014**, *11*, 70. [CrossRef] [PubMed]
95. de Jong, M.D.; Simmons, C.P.; Thanh, T.T.; Hien, V.M.; Smith, G.J.; Chau, T.N.; Hoang, D.M.; Chau, N.V.; Khanh, T.H.; Dong, V.C.; et al. Fatal outcome of human influenza A (H5N1) is associated with high viral load and hypercytokinemia. *Nat. Med.* **2006**, *12*, 1203–1207. [CrossRef] [PubMed]
96. Roosenhoff, R.; Reed, V.; Kenwright, A.; Schutten, M.; Boucher, C.A.; Monto, A.; Clinch, B.; Kumar, D.; Whitley, R.; Nguyen-Van-Tam, J.S.; et al. Viral Kinetics and Resistance Development in Children Treated with Neuraminidase Inhibitors: The Influenza Resistance Information Study (IRIS). *Clin. Infect. Dis.* **2019**, ciz939. [CrossRef] [PubMed]
97. Beigel, J.H.; Aga, E.; Elie-Turenne, M.C.; Cho, J.; Tebas, P.; Clark, C.L.; Metcalf, J.P.; Ozment, C.; Raviprakash, K.; Beeler, J.; et al. Anti-influenza immune plasma for the treatment of patients with severe influenza A: A randomised, double-blind, phase 3 trial. *Lancet Respir. Med.* **2019**, *7*, 941–950. [CrossRef]
98. Sedeyn, K.; Saelens, X. New antibody-based prevention and treatment options for influenza. *Antivir. Res* **2019**, *170*, 104562. [CrossRef]
99. Vermillion, M.S.; Ursin, R.L.; Kuok, D.I.T.; Vom Steeg, L.G.; Wohlgemuth, N.; Hall, O.J.; Fink, A.L.; Sasse, E.; Nelson, A.; Ndeh, R.; et al. Production of amphiregulin and recovery from influenza is greater in males than females. *Biol. Sex Differ.* **2018**, *9*, 24. [CrossRef]
100. Sun, J.; Madan, R.; Karp, C.L.; Braciale, T.J. Effector T cells control lung inflammation during acute influenza virus infection by producing IL-10. *Nat. Med.* **2009**, *15*, 277–284. [CrossRef]
101. Tate, M.D.; Schilter, H.C.; Brooks, A.G.; Reading, P.C. Responses of mouse airway epithelial cells and alveolar macrophages to virulent and avirulent strains of influenza A virus. *Viral Immunol.* **2011**, *24*, 77–88. [CrossRef]
102. Engeland, C.G.; Bosch, J.A.; Cacioppo, J.T.; Marucha, P.T. Mucosal wound healing: The roles of age and sex. *Arch. Surg.* **2006**, *141*, 1193–1197. [CrossRef] [PubMed]
103. Bandyopadhyay, G.; Huyck, H.L.; Misra, R.S.; Bhattacharya, S.; Wang, Q.; Mereness, J.; Lillis, J.; Myers, J.R.; Ashton, J.; Bushnell, T.; et al. Dissociation, cellular isolation, and initial molecular characterization of neonatal and pediatric human lung tissues. *Am. J. Physiol. Lung Cell. Mol. Physiol.* **2018**, *315*, L576–L583. [CrossRef] [PubMed]
104. Kyle, J.E.; Clair, G.; Bandyopadhyay, G.; Misra, R.S.; Zink, E.M.; Bloodsworth, K.J.; Shukla, A.K.; Du, Y.; Lillis, J.; Myers, J.R.; et al. Cell type-resolved human lung lipidome reveals cellular cooperation in lung function. *Sci. Rep.* **2018**, *8*, 13455. [CrossRef] [PubMed]
105. Du, Y.; Clair, G.C.; Al Alam, D.; Danopoulos, S.; Schnell, D.; Kitzmiller, J.A.; Misra, R.S.; Bhattacharya, S.; Warburton, D.; Mariani, T.J.; et al. Integration of transcriptomic and proteomic data identifies biological functions in cell populations from human infant lung. *Am. J. Physiol. Lung Cell. Mol. Physiol.* **2019**, *317*, L347–L360. [CrossRef] [PubMed]
106. Simon, A.K.; Hollander, G.A.; McMichael, A. Evolution of the immune system in humans from infancy to old age. *Proc. Biol. Sci.* **2015**, *282*, 20143085. [CrossRef] [PubMed]
107. Esteve-Sole, A.; Luo, Y.; Vlagea, A.; Deya-Martinez, A.; Yague, J.; Plaza-Martin, A.M.; Juan, M.; Alsina, L. B Regulatory Cells: Players in Pregnancy and Early Life. *Int. J. Mol. Sci.* **2018**, *19*, 2099. [CrossRef] [PubMed]

108. Tosato, F.; Bucciol, G.; Pantano, G.; Putti, M.C.; Sanzari, M.C.; Basso, G.; Plebani, M. Lymphocytes subsets reference values in childhood. *Cytometry A* **2015**, *87*, 81–85. [CrossRef]
109. Wang, G.; Miyahara, Y.; Guo, Z.; Khattar, M.; Stepkowski, S.M.; Chen, W. "Default" generation of neonatal regulatory T cells. *J. Immunol.* **2010**, *185*, 71–78. [CrossRef]
110. Takahata, Y.; Nomura, A.; Takada, H.; Ohga, S.; Furuno, K.; Hikino, S.; Nakayama, H.; Sakaguchi, S.; Hara, T. CD25+CD4+ T cells in human cord blood: An immunoregulatory subset with naive phenotype and specific expression of forkhead box p3 (Foxp3) gene. *Exp. Hematol.* **2004**, *32*, 622–629. [CrossRef]
111. Sarvaria, A.; Basar, R.; Mehta, R.S.; Shaim, H.; Muftuoglu, M.; Khoder, A.; Sekine, T.; Gokdemir, E.; Kondo, K.; Marin, D.; et al. IL-10+ regulatory B cells are enriched in cord blood and may protect against cGVHD after cord blood transplantation. *Blood* **2016**, *128*, 1346–1361. [CrossRef]
112. Esteve-Sole, A.; Teixido, I.; Deya-Martinez, A.; Yague, J.; Plaza-Martin, A.M.; Juan, M.; Alsina, L. Characterization of the Highly Prevalent Regulatory CD24(hi)CD38(hi) B-Cell Population in Human Cord Blood. *Front. Immunol.* **2017**, *8*, 201. [CrossRef] [PubMed]
113. Liu, P.; Li, L.; Fan, P.; Zheng, J.; Zhao, D. High-dose of intravenous immunoglobulin modulates immune tolerance in premature infants. *BMC Pediatr.* **2018**, *18*, 74. [CrossRef] [PubMed]
114. Rueda, C.M.; Moreno-Fernandez, M.E.; Jackson, C.M.; Kallapur, S.G.; Jobe, A.H.; Chougnet, C.A. Neonatal regulatory T cells have reduced capacity to suppress dendritic cell function. *Eur. J. Immunol.* **2015**, *45*, 2582–2592. [CrossRef] [PubMed]
115. Coates, B.M.; Staricha, K.L.; Wiese, K.M.; Ridge, K.M. Influenza A Virus Infection, Innate Immunity, and Childhood. *JAMA Pediatr.* **2015**, *169*, 956–963. [CrossRef] [PubMed]
116. Papaioannou, N.E.; Pasztoi, M.; Schraml, B.U. Understanding the Functional Properties of Neonatal Dendritic Cells: A Doorway to Enhance Vaccine Effectiveness? *Front. Immunol.* **2018**, *9*, 3123. [CrossRef] [PubMed]
117. Lawrence, S.M.; Corriden, R.; Nizet, V. Age-Appropriate Functions and Dysfunctions of the Neonatal Neutrophil. *Front. Pediatr.* **2017**, *5*, 23. [CrossRef]
118. Kollmann, T.R.; Kampmann, B.; Mazmanian, S.K.; Marchant, A.; Levy, O. Protecting the Newborn and Young Infant from Infectious Diseases: Lessons from Immune Ontogeny. *Immunity* **2017**, *46*, 350–363. [CrossRef]
119. Wisgrill, L.; Groschopf, A.; Herndl, E.; Sadeghi, K.; Spittler, A.; Berger, A.; Forster-Waldl, E. Reduced TNF-alpha response in preterm neonates is associated with impaired nonclassic monocyte function. *J. Leukoc. Biol.* **2016**, *100*, 607–612. [CrossRef]
120. de Jong, E.; Strunk, T.; Burgner, D.; Lavoie, P.M.; Currie, A. The phenotype and function of preterm infant monocytes: Implications for susceptibility to infection. *J. Leukoc. Biol.* **2017**, *102*, 645–656. [CrossRef]
121. Collins, A.; Weitkamp, J.H.; Wynn, J.L. Why are preterm newborns at increased risk of infection? Archives of disease in childhood. *Fetal Neonatal Ed.* **2018**, *103*, F391–F394. [CrossRef]
122. D'Angio, C.T.; Wyman, C.P.; Misra, R.S.; Halliley, J.L.; Wang, H.; Hunn, J.E.; Fallone, C.M.; Lee, F.E. Plasma cell and serum antibody responses to influenza vaccine in preterm and full-term infants. *Vaccine* **2017**, *35*, 5163–5171. [CrossRef] [PubMed]
123. D'Angio, C.T.; Heyne, R.J.; Duara, S.; Holmes, L.C.; O'Shea, T.M.; Wang, H.; Wang, D.; Sanchez, P.J.; Welliver, R.C.; Ryan, R.M.; et al. Immunogenicity of trivalent influenza vaccine in extremely low-birth-weight, premature versus term infants. *Pediatr. Infect. Dis. J.* **2011**, *30*, 570–574. [CrossRef] [PubMed]
124. Misra, R.; Shah, S.; Fowell, D.; Wang, H.; Scheible, K.; Misra, S.; Huyck, H.; Wyman, C.; Ryan, R.M.; Reynolds, A.M.; et al. Preterm cord blood CD4(+) T cells exhibit increased IL-6 production in chorioamnionitis and decreased CD4(+) T cells in bronchopulmonary dysplasia. *Hum. Immunol.* **2015**, *76*, 329–338. [CrossRef] [PubMed]
125. Jobe, A.H. Mechanisms of Lung Injury and Bronchopulmonary Dysplasia. *Am. J. Perinatol.* **2016**, *33*, 1076–1078. [CrossRef]
126. Mitra, S.; Disher, T.; Pichler, G.; D'Souza, B.; McCord, H.; Chayapathi, V.; Jones, K.; Schmolzer, G. Delivery room interventions to prevent bronchopulmonary dysplasia in preterm infants: A protocol for a systematic review and network meta-analysis. *BMJ Open* **2019**, *9*, e028066. [CrossRef]
127. Coalson, J.J. Pathology of bronchopulmonary dysplasia. *Semin. Perinatol.* **2006**, *30*, 179–184. [CrossRef]
128. Papagianis, P.C.; Pillow, J.J.; Moss, T.J. Bronchopulmonary dysplasia: Pathophysiology and potential anti-inflammatory therapies. *Paediatr. Respir. Rev.* **2019**, *30*, 34–41. [CrossRef]

129. Collins, J.J.P.; Tibboel, D.; de Kleer, I.M.; Reiss, I.K.M.; Rottier, R.J. The Future of Bronchopulmonary Dysplasia: Emerging Pathophysiological Concepts and Potential New Avenues of Treatment. *Front. Med.* **2017**, *4*, 61. [CrossRef]
130. Bhandari, A.; Carroll, C.; Bhandari, V. BPD Following Preterm Birth: A Model for Chronic Lung Disease and a Substrate for ARDS in Childhood. *Front. Pediatr.* **2016**, *4*, 60. [CrossRef]
131. Homaira, N.; Briggs, N.; Oei, J.L.; Hilder, L.; Bajuk, B.; Snelling, T.; Chambers, G.M.; Jaffe, A. Impact of influenza on hospitalization rates in children with a range of chronic lung diseases. *Influenza Other Respir. Viruses* **2019**, *13*, 233–239. [CrossRef]
132. Black, S.; Nicolay, U.; Vesikari, T.; Knuf, M.; Del Giudice, G.; Della Cioppa, G.; Tsai, T.; Clemens, R.; Rappuoli, R. Hemagglutination inhibition antibody titers as a correlate of protection for inactivated influenza vaccines in children. *Pediatr. Infect. Dis. J.* **2011**, *30*, 1081–1085. [CrossRef] [PubMed]
133. Wilkinson, T.M.; Li, C.K.; Chui, C.S.; Huang, A.K.; Perkins, M.; Liebner, J.C.; Lambkin-Williams, R.; Gilbert, A.; Oxford, J.; Nicholas, B.; et al. Preexisting influenza-specific CD4(+) T cells correlate with disease protection against influenza challenge in humans. *Nat. Med.* **2012**, *18*, 274–280. [CrossRef] [PubMed]
134. Sridhar, S.; Begom, S.; Bermingham, A.; Hoschler, K.; Adamson, W.; Carman, W.; Bean, T.; Barclay, W.; Deeks, J.J.; Lalvani, A. Cellular immune correlates of protection against symptomatic pandemic influenza. *Nat. Med.* **2013**, *19*, 1305–1312. [CrossRef] [PubMed]
135. Belshe, R.B.; Edwards, K.M.; Vesikari, T.; Black, S.V.; Walker, R.E.; Hultquist, M.; Kemble, G.; Connor, E.M.; Group, C.-T.C.E.S. Live attenuated versus inactivated influenza vaccine in infants and young children. *N. Engl. J. Med.* **2007**, *356*, 685–696. [CrossRef] [PubMed]
136. Grohskopf, L.A.; Alyanak, E.; Broder, K.R.; Walter, E.B.; Fry, A.M.; Jernigan, D.B. Prevention and Control of Seasonal Influenza with Vaccines: Recommendations of the Advisory Committee on Immunization Practices-United States, 2019–2020 Influenza Season. *MMWR Recomm. Rep.* **2019**, *68*, 1–21. [CrossRef]
137. ACOG Committee on Obstetric Practice. ACOG Committee Opinion No. 732: Influenza Vaccination During Pregnancy. *Obstet. Gynecol.* **2018**, *131*, e109–e114. [CrossRef]
138. Gagneur, A.; Pinquier, D.; Quach, C. Immunization of preterm infants. *Hum. Vaccines Immunother.* **2015**, *11*, 2556–2563. [CrossRef]
139. Center for Disease Control and Prevention. Special Situations; General Best Practice Guidelines for Immunization: Best Practices Guidance of the Advisory Committee on Immunization Practices (ACIP). Available online: https://www.cdc.gov/vaccines/hcp/acip-recs/general-recs/special-situations.html (accessed on 17 September 2019).
140. Ortiz, J.R.; Perut, M.; Dumolard, L.; Wijesinghe, P.R.; Jorgensen, P.; Ropero, A.M.; Danovaro-Holliday, M.C.; Heffelfinger, J.D.; Tevi-Benissan, C.; Teleb, N.A.; et al. A global review of national influenza immunization policies: Analysis of the 2014 WHO/UNICEF Joint Reporting Form on immunization. *Vaccine* **2016**, *34*, 5400–5405. [CrossRef]
141. Paules, C.; Subbarao, K. Influenza. *Lancet* **2017**, *390*, 697–708. [CrossRef]
142. Rajao, D.S.; Gauger, P.C.; Anderson, T.K.; Lewis, N.S.; Abente, E.J.; Killian, M.L.; Perez, D.R.; Sutton, T.C.; Zhang, J.; Vincent, A.L. Novel Reassortant Human-Like H3N2 and H3N1 Influenza A Viruses Detected in Pigs Are Virulent and Antigenically Distinct from Swine Viruses Endemic to the United States. *J. Virol.* **2015**, *89*, 11213–11222. [CrossRef]
143. Zhang, W.; Hirve, S.; Kieny, M.P. Seasonal vaccines—Critical path to pandemic influenza response. *Vaccine* **2017**, *35*, 851–852. [CrossRef] [PubMed]
144. Sheikh, S.; Biundo, E.; Courcier, S.; Damm, O.; Launay, O.; Maes, E.; Marcos, C.; Matthews, S.; Meijer, C.; Poscia, A.; et al. A report on the status of vaccination in Europe. *Vaccine* **2018**, *36*, 4979–4992. [CrossRef] [PubMed]
145. Ohlrogge, A.W.; Suggs, L.S. Flu vaccination communication in Europe: What does the government communicate and how? *Vaccine* **2018**, *36*, 6512–6519. [CrossRef] [PubMed]
146. Samaan, G.; McPherson, M.; Partridge, J. A review of the evidence to support influenza vaccine introduction in countries and areas of WHO's Western Pacific Region. *PLoS ONE* **2013**, *8*, e70003. [CrossRef]
147. Fischer, W.A., 2nd; Gong, M.; Bhagwanjee, S.; Sevransky, J. Global burden of influenza as a cause of cardiopulmonary morbidity and mortality. *Glob. Heart* **2014**, *9*, 325–336. [CrossRef]

148. Lu, P.J.; Hung, M.C.; O'Halloran, A.C.; Ding, H.; Srivastav, A.; Williams, W.W.; Singleton, J.A. Seasonal Influenza Vaccination Coverage Trends Among Adult Populations, U.S., 2010–2016. *Am. J. Prev. Med.* **2019**, *57*, 458–469. [CrossRef]
149. Buchan, S.A.; Kwong, J.C. Trends in influenza vaccine coverage and vaccine hesitancy in Canada, 2006/07 to 2013/14: Results from cross-sectional survey data. *CMAJ Open* **2016**, *4*, E455–E462. [CrossRef]
150. Ding, H.; Kahn, K.E.; Black, C.L.; O'Halloran, A.; Lu, P.J.; Williams, W.W. Influenza Vaccination Coverage Among Pregnant Women in the U.S., 2012–2015. *Am. J. Prev. Med.* **2019**, *56*, 477–486. [CrossRef]
151. Flannery, B.; Reynolds, S.B.; Blanton, L.; Santibanez, T.A.; O'Halloran, A.; Lu, P.-J.; Chen, J.; Foppa, I.M.; Gargiullo, P.; Bresee, J.; et al. Influenza Vaccine Effectiveness Against Pediatric Deaths: 2010–2014. *Pediatrics* **2017**, *139*, e20164244. [CrossRef]
152. Hartel, C.; Humberg, A.; Viemann, D.; Stein, A.; Orlikowsky, T.; Rupp, J.; Kopp, M.V.; Herting, E.; Gopel, W. Preterm Birth during Influenza Season Is Associated with Adverse Outcome in Very Low Birth Weight Infants. *Front. Pediatrics* **2016**, *4*, 130. [CrossRef]
153. Neuzil, K.M.; Mellen, B.G.; Wright, P.F.; Mitchel, E.F., Jr.; Griffin, M.R. The effect of influenza on hospitalizations, outpatient visits, and courses of antibiotics in children. *N. Engl. J. Med.* **2000**, *342*, 225–231. [CrossRef] [PubMed]
154. Hofstetter, A.M.; Jacobson, E.N.; deHart, M.P.; Englund, J.A. Early Childhood Vaccination Status of Preterm Infants. *Pediatrics* **2019**, *144*, e20183520. [CrossRef] [PubMed]
155. World Health Organization. Vaccines against influenza WHO position paper—November 2012. *Wkly. Epidemiol. Rec.* **2012**, *87*, 461–476.
156. Mertz, D.; Lo, C.K.; Lytvyn, L.; Ortiz, J.R.; Loeb, M.; Flurisk, I. Pregnancy as a risk factor for severe influenza infection: An individual participant data meta-analysis. *BMC Infect. Dis.* **2019**, *19*, 683. [CrossRef] [PubMed]
157. Mertz, D.; Kim, T.H.; Johnstone, J.; Lam, P.P.; Science, M.; Kuster, S.P.; Fadel, S.A.; Tran, D.; Fernandez, E.; Bhatnagar, N.; et al. Populations at risk for severe or complicated influenza illness: Systematic review and meta-analysis. *BMJ* **2013**, *347*, f5061. [CrossRef]
158. Mertz, D.; Geraci, J.; Winkup, J.; Gessner, B.D.; Ortiz, J.R.; Loeb, M. Pregnancy as a risk factor for severe outcomes from influenza virus infection: A systematic review and meta-analysis of observational studies. *Vaccine* **2017**, *35*, 521–528. [CrossRef]
159. Sperling, R.S.; Riley, L.E.; Immunization and Emerging Infections Expert Work Group. Influenza Vaccination, Pregnancy Safety, and Risk of Early Pregnancy Loss. *Obstet. Gynecol.* **2018**, *131*, 799–802. [CrossRef]
160. Zerbo, O.; Modaressi, S.; Chan, B.; Goddard, K.; Lewis, N.; Bok, K.; Fireman, B.; Klein, N.P.; Baxter, R. No association between influenza vaccination during pregnancy and adverse birth outcomes. *Vaccine* **2017**, *35*, 3186–3190. [CrossRef]
161. Walsh, L.K.; Donelle, J.; Dodds, L.; Hawken, S.; Wilson, K.; Benchimol, E.I.; Chakraborty, P.; Guttmann, A.; Kwong, J.C.; MacDonald, N.E.; et al. Health outcomes of young children born to mothers who received 2009 pandemic H1N1 influenza vaccination during pregnancy: Retrospective cohort study. *BMJ* **2019**, *366*, l4151. [CrossRef]
162. McHugh, L.; Marshall, H.S.; Perrett, K.P.; Nolan, T.; Wood, N.; Lambert, S.B.; Richmond, P.; Ware, R.S.; Binks, P.; Binks, M.J.; et al. The Safety of Influenza and Pertussis Vaccination in Pregnancy in a Cohort of Australian Mother-Infant Pairs, 2012–2015: FluMum Study. *Clin. Infect. Dis.* **2019**, *68*, 402–408. [CrossRef]
163. Chambers, C.D.; Johnson, D.L.; Xu, R.; Luo, Y.J.; Louik, C.; Mitchell, A.A.; Schatz, M.; Jones, K.L.; OTIS Collaborative Research Group. Safety of the 2010-11, 2011–2012, 2012–2013, and 2013-14 seasonal influenza vaccines in pregnancy: Birth defects, spontaneous abortion, preterm delivery, and small for gestational age infants, a study from the cohort arm of VAMPSS. *Vaccine* **2016**, *34*, 4443–4449. [CrossRef] [PubMed]
164. Moro, P.; Baumblatt, J.; Lewis, P.; Cragan, J.; Tepper, N.; Cano, M. Surveillance of Adverse Events After Seasonal Influenza Vaccination in Pregnant Women and Their Infants in the Vaccine Adverse Event Reporting System, July 2010–May 2016. *Drug Saf.* **2017**, *40*, 145–152. [CrossRef]
165. Eaton, A.; Lewis, N.; Fireman, B.; Hansen, J.; Baxter, R.; Gee, J.; Klein, N.P. Birth outcomes following immunization of pregnant women with pandemic H1N1 influenza vaccine 2009–2010. *Vaccine* **2018**, *36*, 2733–2739. [CrossRef] [PubMed]
166. Nunes, M.C.; Aqil, A.R.; Omer, S.B.; Madhi, S.A. The Effects of Influenza Vaccination during Pregnancy on Birth Outcomes: A Systematic Review and Meta-Analysis. *Am. J. Perinatol.* **2016**, *33*, 1104–1114. [CrossRef] [PubMed]

167. Hutcheon, J.A.; Fell, D.B.; Jackson, M.L.; Kramer, M.S.; Ortiz, J.R.; Savitz, D.A.; Platt, R.W. Detectable Risks in Studies of the Fetal Benefits of Maternal Influenza Vaccination. *Am. J. Epidemiol.* **2016**, *184*, 227–232. [CrossRef] [PubMed]
168. Zhang, C.; Wang, X.; Liu, D.; Zhang, L.; Sun, X. A systematic review and meta-analysis of fetal outcomes following the administration of influenza A/H1N1 vaccination during pregnancy. *Int. J. Gynaecol. Obstet.* **2018**, *141*, 141–150. [CrossRef] [PubMed]
169. Kallen, B.; Olausson, P.O. Vaccination against H1N1 influenza with Pandemrix((R)) during pregnancy and delivery outcome: A Swedish register study. *BJOG* **2012**, *119*, 1583–1590. [CrossRef] [PubMed]
170. van der Maas, N.; Dijs-Elsinga, J.; Kemmeren, J.; van Lier, A.; Knol, M.; de Melker, H. Safety of vaccination against influenza A (H1N1) during pregnancy in the Netherlands: Results on pregnancy outcomes and infant's health: Cross-sectional linkage study. *BJOG* **2016**, *123*, 709–717. [CrossRef]
171. He, J.; Liu, Z.W.; Lu, Y.P.; Li, T.Y.; Liang, X.J.; Arck, P.C.; Huang, S.M.; Hocher, B.; Chen, Y.P. A Systematic Review and Meta-Analysis of Influenza A Virus Infection During Pregnancy Associated with an Increased Risk for Stillbirth and Low Birth Weight. *Kidney Blood Press. Res.* **2017**, *42*, 232–243. [CrossRef]
172. Ribeiro, A.F.; Pellini, A.C.G.; Kitagawa, B.Y.; Marques, D.; Madalosso, G.; Fred, J.; Albernaz, R.K.M.; Carvalhanas, T.; Zanetta, D.M.T. Severe influenza A(H1N1)pdm09 in pregnant women and neonatal outcomes, State of Sao Paulo, Brazil, 2009. *PLoS ONE* **2018**, *13*, e0194392. [CrossRef]
173. Newsome, K.; Alverson, C.J.; Williams, J.; McIntyre, A.F.; Fine, A.D.; Wasserman, C.; Lofy, K.H.; Acosta, M.; Louie, J.K.; Jones-Vessey, K.; et al. Outcomes of infants born to women with influenza A(H1N1)pdm09. *Birth Defects Res.* **2019**, *111*, 88–95. [CrossRef]
174. Fell, D.B.; Savitz, D.A.; Kramer, M.S.; Gessner, B.D.; Katz, M.A.; Knight, M.; Luteijn, J.M.; Marshall, H.; Bhat, N.; Gravett, M.G.; et al. Maternal influenza and birth outcomes: Systematic review of comparative studies. *BJOG* **2017**, *124*, 48–59. [CrossRef]
175. Madhi, S.A.; Cutland, C.L.; Kuwanda, L.; Weinberg, A.; Hugo, A.; Jones, S.; Adrian, P.V.; Niekerk, N.V.; Treurnicht, F.; Ortiz, J.R.; et al. Influenza Vaccination of Pregnant Women and Protection of Their Infants. *N. Engl. J. Med.* **2014**, *371*, 918–931. [CrossRef]
176. Nunes, M.C.; Cutland, C.L.; Dighero, B.; Bate, J.; Jones, S.; Hugo, A.; van Niekerk, N.; Kuwanda, L.; Izu, A.; Weinberg, A.; et al. Kinetics of Hemagglutination-Inhibiting Antibodies Following Maternal Influenza Vaccination Among Mothers With and Those Without HIV Infection and Their Infants. *J. Infect. Dis.* **2015**, *212*, 1976–1987. [CrossRef]
177. Malek, A.; Sager, R.; Kuhn, P.; Nicolaides, K.H.; Schneider, H. Evolution of maternofetal transport of immunoglobulins during human pregnancy. *Am. J. Reprod. Immunol.* **1996**, *36*, 248–255. [CrossRef]
178. Palmeira, P.; Costa-Carvalho, B.T.; Arslanian, C.; Pontes, G.N.; Nagao, A.T.; Carneiro-Sampaio, M.M. Transfer of antibodies across the placenta and in breast milk from mothers on intravenous immunoglobulin. *Pediatric Allergy Immunol.* **2009**, *20*, 528–535. [CrossRef]
179. Nunes, M.C.; Cutland, C.L.; Jones, S.; Downs, S.; Weinberg, A.; Ortiz, J.R.; Neuzil, K.M.; Simões, E.A.F.; Klugman, K.P.; Madhi, S.A. Efficacy of Maternal Influenza Vaccination Against All-Cause Lower Respiratory Tract Infection Hospitalizations in Young Infants: Results From a Randomized Controlled Trial. *Clin. Infect. Dis.* **2017**, *65*, 1066–1071. [CrossRef]
180. Zaman, K.; Roy, E.; Arifeen, S.E.; Rahman, M.; Raqib, R.; Wilson, E.; Omer, S.B.; Shahid, N.S.; Breiman, R.F.; Breiman, R.E.; et al. Effectiveness of Maternal Influenza Immunization in Mothers and Infants. *N. Engl. J. Med.* **2008**, *359*, 1555–1564. [CrossRef]
181. Benowitz, I.; Esposito, D.B.; Gracey, K.D.; Shapiro, E.D.; Vázquez, M. Influenza Vaccine Given to Pregnant Women Reduces Hospitalization Due to Influenza in Their Infants. *Clin. Infect. Dis.* **2010**, *51*, 1355–1361. [CrossRef]
182. Zhong, Z.; Haltalli, M.; Holder, B.; Rice, T.; Donaldson, B.; O'Driscoll, M.; Le-Doare, K.; Kampmann, B.; Tregoning, J.S. The impact of timing of maternal influenza immunization on infant antibody levels at birth. *Clin. Exp. Immunol.* **2019**, *195*, 139–152. [CrossRef]
183. Schlaudecker, E.P.; Steinhoff, M.C.; Omer, S.B.; McNeal, M.M.; Roy, E.; Arifeen, S.E.; Dodd, C.N.; Raqib, R.; Breiman, R.F.; Zaman, K. IgA and neutralizing antibodies to influenza a virus in human milk: A randomized trial of antenatal influenza immunization. *PLoS ONE* **2013**, *8*, e70867. [CrossRef]
184. Jarvinen, K.M.; Wang, J.; Seppo, A.E.; Zand, M. Novel multiplex assay for profiling influenza antibodies in breast milk and serum of mother-infant pairs. *F1000Research* **2018**, *7*, 1822. [CrossRef]

185. Brady, R.C.; Jackson, L.A.; Frey, S.E.; Shane, A.L.; Walter, E.B.; Swamy, G.K.; Schlaudecker, E.P.; Szefer, E.; Wolff, M.; McNeal, M.M.; et al. Randomized trial comparing the safety and antibody responses to live attenuated versus inactivated influenza vaccine when administered to breastfeeding women. *Vaccine* **2018**, *36*, 4663–4671. [CrossRef]

186. Demers-Mathieu, V.; Underwood, M.A.; Beverly, R.L.; Nielsen, S.D.; Dallas, D.C. Comparison of Human Milk Immunoglobulin Survival during Gastric Digestion between Preterm and Term Infants. *Nutrients* **2018**, *10*, 631. [CrossRef]

187. van den Berg, J.P.; Westerbeek, E.A.; Berbers, G.A.; van Gageldonk, P.G.; van der Klis, F.R.; van Elburg, R.M. Transplacental transport of IgG antibodies specific for pertussis, diphtheria, tetanus, haemophilus influenzae type b, and Neisseria meningitidis serogroup C is lower in preterm compared with term infants. *Pediatric Infect. Dis. J.* **2010**, *29*, 801–805. [CrossRef]

188. Demers-Mathieu, V.; Huston, R.K.; Markell, A.M.; McCulley, E.A.; Martin, R.L.; Dallas, D.C. Antenatal Influenza A-Specific IgA, IgM, and IgG Antibodies in Mother's Own Breast Milk and Donor Breast Milk, and Gastric Contents and Stools from Preterm Infants. *Nutrients* **2019**, *11*, 1567. [CrossRef]

189. Sunwoo, S.Y.; Schotsaert, M.; Morozov, I.; Davis, A.S.; Li, Y.; Lee, J.; McDowell, C.; Meade, P.; Nachbagauer, R.; Garcia-Sastre, A.; et al. A Universal Influenza Virus Vaccine Candidate Tested in a Pig Vaccination-Infection Model in the Presence of Maternal Antibodies. *Vaccines* **2018**, *6*, 64. [CrossRef]

190. Grabarics, M.; Csernak, O.; Balogh, R.; Beni, S. Analytical characterization of human milk oligosaccharides-potential applications in pharmaceutical analysis. *J. Pharm. Biomed. Anal.* **2017**, *146*, 168–178. [CrossRef]

191. Trend, S.; Strunk, T.; Lloyd, M.L.; Kok, C.H.; Metcalfe, J.; Geddes, D.T.; Lai, C.T.; Richmond, P.; Doherty, D.A.; Simmer, K.; et al. Levels of innate immune factors in preterm and term mothers' breast milk during the 1st month postpartum. *Br. J. Nutr.* **2016**, *115*, 1178–1193. [CrossRef]

192. Bedin, A.S.; Moles, J.P.; Rutagwera, D.; Nagot, N.; Kankasa, C.; Tylleskar, T.; Valverde-Villegas, J.M.; Durand, M.; Van de Perre, P.; Tuaillon, E. MAIT cells, TCR gammadelta+ cells and ILCs cells in human breast milk and blood from HIV infected and uninfected women. *Pediatric Allergy Immunol.* **2019**, *30*, 479–487. [CrossRef]

193. Xiao, L.; Leusink-Muis, T.; Kettelarij, N.; van Ark, I.; Blijenberg, B.; Hesen, N.A.; Stahl, B.; Overbeek, S.A.; Garssen, J.; Folkerts, G.; et al. Human Milk Oligosaccharide 2′-Fucosyllactose Improves Innate and Adaptive Immunity in an Influenza-Specific Murine Vaccination Model. *Front. Immunol.* **2018**, *9*, 452. [CrossRef]

194. Coulet, M.; Phothirath, P.; Allais, L.; Schilter, B. Pre-clinical safety evaluation of the synthetic human milk, nature-identical, oligosaccharide 2′-O-Fucosyllactose (2′FL). *Regul. Toxicol. Pharmacol.* **2014**, *68*, 59–69. [CrossRef]

195. Vazquez, E.; Barranco, A.; Ramirez, M.; Gruart, A.; Delgado-Garcia, J.M.; Martinez-Lara, E.; Blanco, S.; Martin, M.J.; Castanys, E.; Buck, R.; et al. Effects of a human milk oligosaccharide, 2′-fucosyllactose, on hippocampal long-term potentiation and learning capabilities in rodents. *J. Nutr. Biochem.* **2015**, *26*, 455–465. [CrossRef]

196. Elison, E.; Vigsnaes, L.K.; Rindom Krogsgaard, L.; Rasmussen, J.; Sorensen, N.; McConnell, B.; Hennet, T.; Sommer, M.O.; Bytzer, P. Oral supplementation of healthy adults with 2′-O-fucosyllactose and lacto-N-neotetraose is well tolerated and shifts the intestinal microbiota. *Br. J. Nutr.* **2016**, *116*, 1356–1368. [CrossRef]

197. Tregoning, J.S.; Russell, R.F.; Kinnear, E. Adjuvanted influenza vaccines. *Hum. Vaccines Immunother.* **2018**, *14*, 550–564. [CrossRef]

198. Diez-Domingo, J.; Garces-Sanchez, M.; Baldo, J.M.; Planelles, M.V.; Ubeda, I.; JuBert, A.; Mares, J.; Moris, P.; Garcia-Corbeira, P.; Drame, M.; et al. Immunogenicity and Safety of H5N1 A/Vietnam/1194/2004 (Clade 1) AS03-adjuvanted prepandemic candidate influenza vaccines in children aged 3 to 9 years: A phase ii, randomized, open, controlled study. *Pediatric Infect. Dis. J.* **2010**, *29*, e35–e46. [CrossRef]

199. Waddington, C.S.; Walker, W.T.; Oeser, C.; Reiner, A.; John, T.; Wilkins, S.; Casey, M.; Eccleston, P.E.; Allen, R.J.; Okike, I.; et al. Safety and immunogenicity of AS03B adjuvanted split virion versus non-adjuvanted whole virion H1N1 influenza vaccine in UK children aged 6 months-12 years: Open label, randomised, parallel group, multicentre study. *BMJ* **2010**, *340*, c2649. [CrossRef]

200. Monto, A.S.; Malosh, R.E.; Petrie, J.G.; Martin, E.T. The Doctrine of Original Antigenic Sin: Separating Good From Evil. *J. Infect. Dis.* **2017**, *215*, 1782–1788. [CrossRef]

201. Jackson, L.A.; Chen, W.H.; Stapleton, J.T.; Dekker, C.L.; Wald, A.; Brady, R.C.; Edupuganti, S.; Winokur, P.; Mulligan, M.J.; Keyserling, H.L.; et al. Immunogenicity and safety of varying dosages of a monovalent 2009 H1N1 influenza vaccine given with and without AS03 adjuvant system in healthy adults and older persons. *J. Infect. Dis.* **2012**, *206*, 811–820. [CrossRef]
202. Carmona, A.; Omenaca, F.; Tejedor, J.C.; Merino, J.M.; Vaman, T.; Dieussaert, I.; Gillard, P.; Aristegui, J. Immunogenicity and safety of AS03-adjuvanted 2009 influenza A H1N1 vaccine in children 6–35 months. *Vaccine* **2010**, *28*, 5837–5844. [CrossRef]
203. Garcia-Sicilia, J.; Gillard, P.; Carmona, A.; Tejedor, J.C.; Aristegui, J.; Merino, J.M.; Behre, U.; Caplanusi, A.; Vaman, T.; Dieussaert, I. Immunogenicity and safety of AS03-adjuvanted H1N1 pandemic vaccines in children and adolescents. *Vaccine* **2011**, *29*, 4353–4361. [CrossRef]
204. Laupeze, B.; Herve, C.; Di Pasquale, A.; Tavares Da Silva, F. Adjuvant Systems for vaccines: 13years of post-licensure experience in diverse populations have progressed the way adjuvanted vaccine safety is investigated and understood. *Vaccine* **2019**, *37*, 5670–5680. [CrossRef]
205. Esposito, S.; Pugni, L.; Daleno, C.; Ronchi, A.; Valzano, A.; Serra, D.; Mosca, F.; Principi, N. Influenza A/H1N1 MF59-adjuvanted vaccine in preterm and term children aged 6 to 23 months. *Pediatrics* **2011**, *127*, e1161–e1168. [CrossRef]
206. Nakaya, H.I.; Clutterbuck, E.; Kazmin, D.; Wang, L.; Cortese, M.; Bosinger, S.E.; Patel, N.B.; Zak, D.E.; Aderem, A.; Dong, T.; et al. Systems biology of immunity to MF59-adjuvanted versus nonadjuvanted trivalent seasonal influenza vaccines in early childhood. *Proc. Natl. Acad. Sci. USA* **2016**, *113*, 1853–1858. [CrossRef]
207. Zedda, L.; Forleo-Neto, E.; Vertruyen, A.; Raes, M.; Marchant, A.; Jansen, W.; Clouting, H.; Arora, A.; Beatty, M.E.; Galli, G.; et al. Dissecting the immune response to MF59-adjuvanted and nonadjuvanted seasonal influenza vaccines in children less than three years of age. *Pediatr. Infect. Dis. J.* **2015**, *34*, 73–78. [CrossRef]
208. Mastelic Gavillet, B.; Eberhardt, C.S.; Auderset, F.; Castellino, F.; Seubert, A.; Tregoning, J.S.; Lambert, P.H.; de Gregorio, E.; Del Giudice, G.; Siegrist, C.A. MF59 Mediates Its B Cell Adjuvanticity by Promoting T Follicular Helper Cells and Thus Germinal Center Responses in Adult and Early Life. *J. Immunol.* **2015**, *194*, 4836–4845. [CrossRef]
209. Haber, P.; Moro, P.L.; Ng, C.; Dores, G.M.; Lewis, P.; Cano, M. Post-licensure surveillance of trivalent adjuvanted influenza vaccine (aIIV3; Fluad), Vaccine Adverse Event Reporting System (VAERS), United States, July 2016–June 2018. *Vaccine* **2019**, *37*, 1516–1520. [CrossRef]
210. Sakala, I.G.; Honda-Okubo, Y.; Fung, J.; Petrovsky, N. Influenza immunization during pregnancy: Benefits for mother and infant. *Hum. Vaccines Immunother.* **2016**, *12*, 3065–3071. [CrossRef]
211. Baum, U.; Leino, T.; Gissler, M.; Kilpi, T.; Jokinen, J. Perinatal survival and health after maternal influenza A(H1N1)pdm09 vaccination: A cohort study of pregnancies stratified by trimester of vaccination. *Vaccine* **2015**, *33*, 4850–4857. [CrossRef]
212. Ludvigsson, J.F.; Strom, P.; Lundholm, C.; Cnattingius, S.; Ekbom, A.; Ortqvist, A.; Feltelius, N.; Granath, F.; Stephansson, O. Maternal vaccination against H1N1 influenza and offspring mortality: Population based cohort study and sibling design. *BMJ* **2015**, *351*, h5585. [CrossRef]
213. Miller, E.; Andrews, N.; Stellitano, L.; Stowe, J.; Winstone, A.M.; Shneerson, J.; Verity, C. Risk of narcolepsy in children and young people receiving AS03 adjuvanted pandemic A/H1N1 2009 influenza vaccine: Retrospective analysis. *BMJ Br. Med. J.* **2013**, *346*, f794. [CrossRef] [PubMed]
214. Sarkanen, T.O.; Alakuijala, A.P.E.; Dauvilliers, Y.A.; Partinen, M.M. Incidence of narcolepsy after H1N1 influenza and vaccinations: Systematic review and meta-analysis. *Sleep Med. Rev.* **2018**, *38*, 177–186. [CrossRef] [PubMed]
215. Verstraeten, T.; Cohet, C.; Santos, G.D.; Ferreira, G.L.C.; Bollaerts, K.; Bauchau, V.; Shinde, V. Pandemrix™ and narcolepsy: A critical appraisal of the observational studies. *Hum. Vaccines Immunother.* **2015**, *12*, 187–193. [CrossRef] [PubMed]
216. McClenathan, B.M.; Edwards, K.M. Vaccine safety: An evolving evidence-based science. *Br. J. Clin. Pharmacol.* **2019**. [CrossRef]
217. Kaczmarek, M.C.; Duong, U.T.; Ware, R.S.; Lambert, S.B.; Kelly, H.A. The risk of fever following one dose of trivalent inactivated influenza vaccine in children aged >/=6 months to <36 months: A comparison of published and unpublished studies. *Vaccine* **2013**, *31*, 5359–5365. [CrossRef]

218. Halsey, N.A.; Talaat, K.R.; Greenbaum, A.; Mensah, E.; Dudley, M.Z.; Proveaux, T.; Salmon, D.A. The safety of influenza vaccines in children: An Institute for Vaccine Safety white paper. *Vaccine* **2015**, *33* (Suppl. 5), F1–F67. [CrossRef]
219. Stone, C.A., Jr.; Rukasin, C.R.F.; Beachkofsky, T.M.; Phillips, E.J. Immune Mediated Adverse Reactions to Vaccines. *Br. J. Clin. Pharmacol.* **2019**. [CrossRef]
220. Iaconelli, J.; Xuan, L.; Karmacharya, R. HDAC6 Modulates Signaling Pathways Relevant to Synaptic Biology and Neuronal Differentiation in Human Stem-Cell-Derived Neurons. *Int. J. Mol. Sci.* **2019**, *20*, 1605. [CrossRef]
221. Shimabukuro, T.T.; Nguyen, M.; Martin, D.; DeStefano, F. Safety monitoring in the Vaccine Adverse Event Reporting System (VAERS). *Vaccine* **2015**, *33*, 4398–4405. [CrossRef]
222. Rizzo, C.; Rezza, G.; Ricciardi, W. Strategies in recommending influenza vaccination in Europe and US. *Hum. Vaccines Immunother.* **2018**, *14*, 693–698. [CrossRef]
223. O'Leary, S.T.; Riley, L.E.; Lindley, M.C.; Allison, M.A.; Albert, A.P.; Fisher, A.; Jiles, A.J.; Crane, L.A.; Hurley, L.P.; Beaty, B.; et al. Obstetrician-Gynecologists' Strategies to Address Vaccine Refusal Among Pregnant Women. *Obstet. Gynecol.* **2019**, *133*, 40–47. [CrossRef] [PubMed]
224. Donzelli, A. Influenza Vaccinations for All Pregnant Women? Better Evidence Is Needed. *Int. J. Environ. Res. Public Health* **2018**, *15*, 2034. [CrossRef] [PubMed]
225. Bresee, J.S.; Lafond, K.E.; McCarron, M.; Azziz-Baumgartner, E.; Chu, S.Y.; Ebama, M.; Hinman, A.R.; Xeuatvongsa, A.; Bino, S.; Richardson, D.; et al. The partnership for influenza vaccine introduction (PIVI): Supporting influenza vaccine program development in low and middle-income countries through public-private partnerships. *Vaccine* **2019**, *37*, 5089–5095. [CrossRef] [PubMed]
226. Ellingson, M.K.; Dudley, M.Z.; Limaye, R.J.; Salmon, D.A.; O'Leary, S.T.; Omer, S.B. Enhancing uptake of influenza maternal vaccine. *Expert Rev. Vaccines* **2019**, *18*, 191–204. [CrossRef] [PubMed]
227. Halder, N.; Kelso, J.K.; Milne, G.J. A model-based economic analysis of pre-pandemic influenza vaccination cost-effectiveness. *BMC Infect. Dis.* **2014**, *14*, 266. [CrossRef] [PubMed]
228. Metcalf, C.J.E.; Ferrari, M.; Graham, A.L.; Grenfell, B.T. Understanding Herd Immunity. *Trends Immunol.* **2015**, *36*, 753–755. [CrossRef]
229. Eichner, M.; Schwehm, M.; Eichner, L.; Gerlier, L. Direct and indirect effects of influenza vaccination. *BMC Infect. Dis.* **2017**, *17*, 308. [CrossRef]
230. Switzer, C.; Babiuk, L.; Loeb, M. Determining optimal community protection strategies for the influenza vaccine. *Expert Rev. Vaccines* **2019**, *18*, 755–764. [CrossRef]
231. National Insititue of Health and Care Excellence. Evidence Reviews for Increasing Uptake in Children. Available online: https://www.nice.org.uk/guidance/ng103/resources/flu-vaccination-increasing-uptake-pdf-66141536272837 (accessed on 20 August 2019).
232. Betsch, C.; Bohm, R. Detrimental effects of introducing partial compulsory vaccination: Experimental evidence. *Eur. J. Public Health* **2016**, *26*, 378–381. [CrossRef]
233. Omer, S.B.; Betsch, C.; Leask, J. Mandate vaccination with care. *Nature* **2019**, *571*, 469–472. [CrossRef]
234. Poland, G.A. MMR vaccine and autism: Vaccine nihilism and postmodern science. *Mayo Clin. Proc.* **2011**, *86*, 869–871. [CrossRef] [PubMed]
235. Retraction-Ileal-lymphoid-nodular hyperplasia, non-specific colitis, and pervasive developmental disorder in children. *Lancet* **2010**, *375*, 445. [CrossRef]
236. Taylor, L.E.; Swerdfeger, A.L.; Eslick, G.D. Vaccines are not associated with autism: An evidence-based meta-analysis of case-control and cohort studies. *Vaccine* **2014**, *32*, 3623–3629. [CrossRef] [PubMed]
237. Zerbo, O.; Modaressi, S.; Goddard, K.; Lewis, E.; Fireman, B.H.; Daley, M.F.; Irving, S.A.; Jackson, L.A.; Donahue, J.G.; Qian, L.; et al. Vaccination Patterns in Children After Autism Spectrum Disorder Diagnosis and in Their Younger Siblings. *JAMA Pediatr.* **2018**, *172*, 469–475. [CrossRef] [PubMed]
238. Vanderslott, S. Study shows lower vaccination rates for younger siblings after autism spectrum disorder diagnosis in older siblings. *Evid. Based Nurs.* **2019**, *22*, 119. [CrossRef]
239. Lane, S.; MacDonald, N.E.; Marti, M.; Dumolard, L. Vaccine hesitancy around the globe: Analysis of three years of WHO/UNICEF Joint Reporting Form data-2015–2017. *Vaccine* **2018**, *36*, 3861–3867. [CrossRef]
240. Nihlen Fahlquist, J. Vaccine hesitancy and trust. Ethical aspects of risk communication. *Scand. J. Public Health* **2018**, *46*, 182–188. [CrossRef]

241. Debellut, F.; Hendrix, N.; Ortiz, J.R.; Lambach, P.; Neuzil, K.M.; Bhat, N.; Pecenka, C. Forecasting demand for maternal influenza immunization in low- and lower-middle-income countries. *PLoS ONE* **2018**, *13*, e0199470. [CrossRef]
242. Launay, O.; Tsatsaris, V. Maternal influenza immunisation in resource-limited settings. *Lancet Infect. Dis.* **2016**, *16*, 983–984. [CrossRef]
243. Tapia, M.D.; Sow, S.O.; Tamboura, B.; Teguete, I.; Pasetti, M.F.; Kodio, M.; Onwuchekwa, U.; Tennant, S.M.; Blackwelder, W.C.; Coulibaly, F.; et al. Maternal immunisation with trivalent inactivated influenza vaccine for prevention of influenza in infants in Mali: A prospective, active-controlled, observer-blind, randomised phase 4 trial. *Lancet Infect. Dis.* **2016**, *16*, 1026–1035. [CrossRef]
244. Breteler, J.K.; Tam, J.S.; Jit, M.; Ket, J.C.; De Boer, M.R. Efficacy and effectiveness of seasonal and pandemic A (H1N1) 2009 influenza vaccines in low and middle income countries: A systematic review and meta-analysis. *Vaccine* **2013**, *31*, 5168–5177. [CrossRef] [PubMed]
245. Ortiz, J.R.; Neuzil, K.M. Influenza immunization of pregnant women in resource-constrained countries: An update for funding and implementation decisions. *Curr. Opin. Infect. Dis.* **2017**, *30*, 455–462. [CrossRef] [PubMed]
246. Global_Alliance_for_Vaccines_and_Immunisation_Alliance. Vaccine Investment Strategy. Available online: https://www.gavi.org/about/strategy/vaccine-investment-strategy/ (accessed on 24 November 2019).
247. Grijalva, C.G.; Zhu, Y.; Williams, D.J.; Self, W.H.; Ampofo, K.; Pavia, A.T.; Stockmann, C.R.; McCullers, J.; Arnold, S.R.; Wunderink, R.G.; et al. Association Between Hospitalization With Community-Acquired Laboratory-Confirmed Influenza Pneumonia and Prior Receipt of Influenza Vaccination. *JAMA* **2015**, *314*, 1488–1497. [CrossRef] [PubMed]
248. Ortiz, J.R.; Neuzil, K.M.; Ahonkhai, V.I.; Gellin, B.G.; Salisbury, D.M.; Read, J.S.; Adegbola, R.A.; Abramson, J.S. Translating vaccine policy into action: A report from the Bill & Melinda Gates Foundation Consultation on the prevention of maternal and early infant influenza in resource-limited settings. *Vaccine* **2012**, *30*, 7134–7140. [CrossRef] [PubMed]

pathogens

MDPI

Review

A Complex Dance: Measuring the Multidimensional Worlds of Influenza Virus Evolution and Anti-Influenza Immune Responses

Jiong Wang [1,*], Alexander Wiltse [1] and Martin S. Zand [1,2]

[1] Department of Medicine, Division of Nephrology, University of Rochester Medical Center, Rochester, NY 14534, USA; Alexander_Wiltse@urmc.rochester.edu (A.W.); Martin_zand@urmc.rochester.edu (M.S.Z.)
[2] Clinical and Translational Science Institute, University of Rochester Medical Center, Rochester, NY 14534, USA
* Correspondence: Jiong_Wang@urmc.rochester.edu

Received: 25 September 2019; Accepted: 12 November 2019; Published: 15 November 2019

Abstract: The human antibody response to influenza virus infection or vaccination is as complicated as it is essential for protection against flu. The constant antigenic changes of the virus to escape human herd immunity hinder the yearly selection of vaccine strains since it is hard to predict which virus strains will circulate for the coming flu season. A "universal" influenza vaccine that could induce broad cross-influenza subtype protection would help to address this issue. However, the human antibody response is intricate and often obscure, with factors such as antigenic seniority or original antigenic sin (OAS), and back-boosting ensuring that each person mounts a unique immune response to infection or vaccination with any new influenza virus strain. Notably, the effects of existing antibodies on cross-protective immunity after repeated vaccinations are unclear. More research is needed to characterize the mechanisms at play, but traditional assays such as hemagglutinin inhibition (HAI) and microneutralization (MN) are excessively limited in scope and too resource-intensive to effectively meet this challenge. In the past ten years, new multiple dimensional assays (MDAs) have been developed to help overcome these problems by simultaneously measuring antibodies against a large panel of influenza hemagglutinin (HA) proteins with a minimal amount of sample in a high throughput way. MDAs will likely be a powerful tool for accelerating the study of the humoral immune response to influenza vaccination and the development of a universal influenza vaccine.

Keywords: influenza virus; humoral response; hemagglutinin (HA) of influenza virus; broad neutralizing antibody(bnAb); heterosubtypic immunity of influenza; original antigenic sin "OAS"; "universal" influenza vaccine; protein microarray assay; mPLEX-Flu assay; multiple dimensional assay (MDA)

1. Introduction

Influenza is a global public health problem, causing approximately 300,000–650,000 global deaths each year [1]. Influenza A and B are the major virus types that infect humans. Antibodies directed against the head domain of the surface glycoprotein hemagglutinin (HA) of influenza virus have proven to be the major source of protective immunity, blocking viral binding to the receptors on the target human cell surface and inhibiting viral entry to target cells. In response to human immunity pressures, antigenically distinct influenza viruses emerge frequently, caused by continual mutation (antigenic drift) [2], or reassortment among viruses from different species (antigenic shift) that can lead to a pandemic with high mortality [3,4].

To date, seasonal influenza vaccines composed of three or four inactivated virus strains are the only licensed vaccines to elicit or boost protective immunity against influenza viruses in the United States. However, both antigenic drift and shift necessitate that the flu vaccine be reformulated and re-administered annually [5]. It is a formidable challenge to select the strains each year to protect against current circulating viruses based on viral surveillance data of the previous year [6], and to produce a large amount of antigenically matched vaccine. Developing a "universal flu vaccine" that induces broadly cross-protective immunity is one strategy to overcome this challenge [7,8].

Antibody mediated immune responses against influenza HA are multi-dimensional, targeting multiple antigenic determinants (epitopes) within the HA molecule. Antibody mediated responses are also incredibly complicated, as they are influenced and altered by an individual's prior influenza exposure history. This includes factors such as "original antigenic sin" (OAS) [9] (also known as HA imprinting [10]) and the shared epitopes between proteins from different influenza strains that induce cross-strain immunity, such as heterosubtypic immunity [11,12]. The effects of pre-existing antibodies on the B cell response to vaccine strains that contain HA antigenic sites similar to those from prior exposures are still unclear. Systems serology, the application of bioinformatics to multidimensional data regarding anti-influenza IgG binding specificity and repertoire in response to vaccination, has emerged as a way to understand these responses, and to aid in vaccine design.

Because of the complex interplay between pre-existing, circulating, anti-HA antibodies and human IgG-mediated influenza responses, the first step in comprehensive analysis is measurement of anti-influenza HA IgG binding patterns against multiple influenza strain HAs. Such measurement is referred to as multi-dimensional, referring to the multiplicity of influenza strain binding reactions quantified. Such measurements are critical for understanding how IgG recognition of shared epitopes across influenza strains can lead to cross-strain protection, and for better defining the functional host anti-HA influenza repertoire.

Various assays exist to measure the host anti-HA influenza antibody response. The assays currently used to estimate the HA IgG antibody binding to single HA proteins, such as hemagglutinin inhibition (HAI) [13,14], micro-neutralization (MN) [15,16] and enzyme-linked immunosorbent assay (ELISA), all require a large amount of serum sample in order to test the cross-reactivity against an array of virus strains. These assays are also expensive and time consuming, limiting their usefulness in unraveling the complexity of cross-reactive antibody patterns to influenza viruses. In contrast, the novel technology of array-based high throughput multiple dimensional assay (MDA) provides a powerful tool to comprehensively analyze the presence and effects of broad cross-reactive antibodies (bcAbs) against the influenza HA protein.

Here, we review the genetic foundations of shared epitopes leading to IgG cross-reactivity between antigenically similar influenza virus strains, and the contributions of these cross-reactivities to OAS and subsequent host immune responses to influenza infection and vaccination. The remainder of the review discusses the technology of MDA, and highlights the application of MDA as a powerful tool for future influenza immunity studies and universal vaccine development.

2. Hemagglutinin (HA) and Its Antibodies

Influenza viruses, the pathogens that cause flu, belong to the *Orthomyxoviridae* family, a group of negative-sense single strain RNA viruses [17]. Influenza type A has two phylogenetic groups based on amino acid sequence and, to date, 18 HA subtypes: Group 1 (H1, H2, H5, H6, H8, H9, H11, H12, H13, H16, HA-like H17, and HA-like H18) and Group 2 (H3, H4, H7, H10, H14, and H15) [18]. Influenza A viruses are further named based on the composition of major surface glycoproteins HA and neuraminidase (NA) (e.g., H1N1 or H3N2). Influenza type B also has two phylogenetically distinct lineages called Yamagata and

Victoria [19]. The major source of human protective immunity is the antibodies directed against the head domain of the HA of influenza virus [20]. HA is the most abundant influenza viral surface glycoprotein and mediates binding to sialic acid expressed on the surface of target host cells. HA is synthesized as a polypeptide (HA0) before being cleaved into HA1 and HA2 subunits, which fold into a trimeric spike. The membrane distal globular head region of HA is composed of HA1 and contains the receptor binding site (RBS) that the virus uses to bind to host cell sialic acid. The stalk region then mediates virus fusion into host cells through structure transformation [21].

Protective antibody-mediated immunity against HA is the first line of defense in preventing influenza virus infection. Such immunity is elicited by prior influenza exposure: infection or vaccination [22,23]. Anti-head HA antibodies typically target epitopes in and around the RBS. Five major B cell epitopes have been identified for H1 (Sa, Sb, Ca1, Ca2, and Cb) [24] and H3 (Eptitopes A–E) influenza strain HAs [25]. The HA head region, formed by HA1, is strongly immunodominant, highly mutable, and strain-specific [26]. The HA stalk region, formed by HA2 as well as the N- and C-terminal ends of HA1 in an alpha-helical structure, supports the head region of HA [27]. The highly conserved nature of the HA stalk makes it a promising target for universal influenza vaccines [28–30].

The goal of universal vaccines is to elicit protective broad cross-reactive antibodies (bcAbs), especially broad neutralizing antibodies (bnAbs). Most head-reactive antibodies are not bcAbs or bnAbs but rather strain-specific. However, more and more head domain recognizing bnAbs have been identified, such as KBm2, 5J8 and CH65, which neutralize a broad spectrum of H1 strain viruses in the MN assay [31–33], and 8M2, which neutralizes many H2 strains [34]. Several head-reactive bnAbs demonstrate heterosubtypic reactivity, such as C05, F045-92 and S139/1, which recognize the conserved receptor binding pocket on the HA head [35–38]. These three bnAbs can neutralize H1, H2, and H9, while C05 can also weakly neutralize the Group 2 H3 influenza virus [35].

Recently, an increasing number of bnAbs have been isolated and identified from the B cell repertoire after influenza virus infection and vaccination [31,39], targeting both head and stalk regions of HA. Some bnAbs that target stalk region of HA neutralize a wider range of influenza types and subtypes [18]. Human monoclonal antibodies CR6261, F10 and A06 were isolated from recently vaccinated donors and shown to neutralize nearly all Group 1 viruses [40–42], while CR8020 and CR8043 neutralize a wide breadth of Group 2 viruses [43,44]. Some monoclonal antibodies show broad cross-group influenza A reactivity, including MEDI8852, 27F3, FI6v3 and CR9114 [45–48]. Notably, there are other in vivo mechanisms involved in antibody mediated broad-protection, such antibody-dependent cell-mediated cytotoxicity (ADCC) and antibody-mediated cellular phagocytosis (ADCP)[49–51]. For example, the novel isolated human mAb FluA-20 was shown to protect mice against lethal challenge with H1, H3, H5 and H7 influenza A subtype viruses [51]. It binds to an extremely well conserved epitope in the peripheral interface of the HA trimer, a novel epitope on the head region of HA, with extra high affinity. After binding to HA, it quickly interferes with the trimeric structure of HA, which blocks viral cell-to-cell spread. This mAb offered protection from influenza virus infection in vivo, but it did not display neutralizing activity in HAI or MN assay in vitro studies [51]. This suggests that there are some bcAbs that are able to protect against influenza virus that would likely be disregarded by traditional assays. BcAbs can be detected by ELISA and other binding assays, including MDAs, thus highlighting the need for more sensitive assays in the search for broad cross-reactive antibodies.

One major issue with the evaluation of bnAb activities by traditional HAI and MN assays is that they can only measure the magnitude of bnAb against specific virus strains, and cannot determine the breadth of bnAbs against a large panel of influenza strains. In addition, during antibody screening, these assays most likely omit bcAbs, such as the human mAb FluA-20 [51], as we discussed above, did not display neutralizing activity in MN assay in vitro.

3. Complexity of Human Immune Responses against Influenza Virus

The complexity of the human immune response to repeated influenza virus exposure is another major obstacle to the development of a universal vaccine. Because of the frequent antigenic drift in circulating influenza strains, humans have more complicated immune responses than can be modeled in naive animals. Each person has a unique history of influenza virus exposure, leading to pre-existing immune repertoires that are activated in the event of an immune challenge with an antigenically similar flu strain.

In 1960, Thomas Francis Jr. reported that antibodies against the first H1N1 flu strain encountered in life would be produced at high levels throughout a person's lifespan, to the detriment of future specific responses to new strains [9]. He coined the term "Original Antigenic Sin" (OAS), now referred to as imprinting, to describe how a specific immune response to a flu strain can be preferentially directed at a previously encountered strain. This phenomenon also relates to the cross-protection provided by pre-existing bcAbs and how the breadth and protective potency of cross-reactive immunity is enhanced by infection or vaccinations. For example, the lower mortality of older individuals during the 2009 H1N1 pandemic is attributed to the structural similarity between the pandemic 2009 "Swine" flu virus HA and the pandemic 1918 "Spanish" flu virus HA, suggesting within-subtype cross-strain protection [52–54]. Importantly, in 2016, using all known fatal human cases of H5N1 and H7N9, Gostic et al. [10] found that childhood H1 and H3 imprinting provided 75% and 80% protection against death from H5N1 and H7N9, respectively. Because H1 and H5 are found in phylogenetic Group 1, and H3 and H7 are found in Group 2, these results suggest that antigenic seniority boosts can offer cross-protection against HA subtypes of the same group [10]. The mechanism of such immune imprinting is unclear, but it has been hypothesized that after a large number of memory B cells (MBCs) are activated during first influenza virus exposure, the next exposure to an influenza strain with some mutated and some shared epitopes will show lower de novo naive memory B cells activation against the *new* epitopes. This suggests that preexisting antibodies could play a role in the MBC response, such as accelerating the clearance of influenza antigens or sterically blocking MBCs from accessing specific epitopes [55,56]. Meanwhile, the MBCs specific for epitopes present in the first strain would proliferate more since they have been activated again [57]. OAS [58], "antigenic seniority" [59] and HA imprinting [10] try to describe the effects of pre-existing antibodies on the antibody response to similar or related influenza virus strains. Due to the multi-dimensional nature of human immune repertoires, single-dimensional assays are extremely limited in their ability to measure the breadth of pre-existing bcAb and MBC responses. MDAs, on the other hand, are an ideal tool for measuring pre-existing bcAb profiles and broad influenza immunity.

4. Multidimensional Assays (MDAs) for Anti-Influenza Antibodies

The gold standard and most widely used assays to evaluate antibody activity against HA and protection in clinical trials are HAI [13,14] and MN [15,16] assays. Both assays are semi-quantitative with a single target virus strain providing a discrete ranked readout of one of 8–14 titer values based on two-fold dilutions of serum samples. Including ELISA, which is less frequently used in influenza studies, all these common methods are single dimensional assays, which require the user to perform antibody testing for each strain of interest separately. This process is not only time-consuming and labor-intensive, but also requires large sample volumes. In addition, these assays are limited in their ability to show the breadth of cross-reactive anti-influenza antibody response.

To overcome the limitations of single dimensional assays, novel multidimensional assays (MDAs) have been developed over the last decade. MDAs are high throughput assays that use protein array technology to simultaneously measure antibodies against a panel of the HA proteins and peptides of multiple influenza virus strains in a single test with minimum amount of sample. They can measure the magnitude and breadth of antibody response against HAs of influenza virus. In general, the purified HA proteins are immobilized on a solid surface such as microchips, membranes or beads, to keep the native structure and provide their maximal binding properties. Then, the reactive antibodies are characterized by binding to the protein, followed by a fluorescent probed secondary antibody that is read by an array chips reader as median fluorescent intensity (MFI). The HA protein or peptide array offers the advantage of multiplex capabilities to generate statistically powerful data while conserving time, money, and requiring minimal sample compared to the traditional assays. While not a functional assay, multiple studies have confirmed that MDA results correlate well with HAI titers [60–64]. Critical for the understanding of OAS, "back-boosting", and the effects of pre-existing cross-strain immunity on current vaccine responses, such methods allow testing reactivity against a large number of antigenically related and disparate influenza proteins (generally HA at the moment).

Based on the immobilizing materials, there are two major types of array-based assays currently used for evaluation of HA antibodies: protein microarray and Luminex assays. The first method involves printing HA protein on chips to estimate the binding antibodies. The first report of HA protein array assay was published in 2010 [65], and since then more than 10 publications have shown its powerful potential to study the breadth of cross-reactivity of HA antibodies on the population level (see the list in Table 1). At present, 283 HA proteins can be printed on one microchip for maximum efficiency [66]. However, this process requires expensive and specialized equipment, including a micro-printer and dedicated scanner.

In contrast to peptide arrays, the Luminex-based MDA method, which involves coupling HA protein to color coded Luminex beads, allows the user increased flexibility to customize the panel by easily combining multiple strain-specific beads without reprinting the entire panel [63,67]. In addition, Luminex readers are more widely available now than chip scanners. However, Luminex-based MDAs support fewer analytes per assay. For example, the Luminex 200 can detect 100 color-coded beads, and the Magpix analyzer can read 50 coded beads [68].

Table 1. Multidimensional assay (MDA) methods for detecting antibodies against influenza HA strains

Methods	Target Antigen	Species	Isotype	Sample Type(s)	Reference
Luminex array	NP, M1 and NS1 proteins	Chicken, turkey	IgY	Serum	[69]
	Whole HA of H1, H3, H5, Flu B	Human	IgA1, IgG1	Serum	[70]
	Whole HA of H1, H3, Flu B	Ferret, mouse, human	IgG, IgA, IgM	Serum	[67]
	Whole HA of H1, H3, H5	Human	IgG	Serum	[71]
	Whole HA of H1, H2, H3, H5, H7, H9, Flu B and chimeric HA	Human	IgG	Serum MBC culture	[63,72–74]
	Whole HA of H1, H2, H3, H5, H7, H9, Flu B	Human	IgG	Purified mAb	[31]
	Whole HA of H1, H2, H3, H5, H7, H9, Flu B and chimeric HA	Human	IgA, IgG	Breast milk Infant serum	[75]
	Whole HA of H1, H2, H3, H5, H7, H9, Flu B and chimeric HA	Human	IgG	Serum	[64,76]
	H1-16 whole HA, and N1-9 whole NA Avian flu	Chicken	IgY	Serum	[77]
Microarray	Random sequence peptides	Human	IgG	Serum	[65]
	Head domain of HA of H1, H2, H3, H5, H7, H9	Human	IgG	Serum or dry blood spots	[60–62,78–81]
	H1-H16 and H18 whole HA protein and/or HA peptides	Human	IgG	Serum	[66,82,83]
	H1-H18 whole HA	Chicken, duck, bat	IgY, IgG	Serum	[79,84,85]
Arrayed Imaging	H1, H3, H6, H5	Human	IgG	Serum	[86]
Reflectometry (AIR)	H1-H12 and Flu B	Mallard duck	IgY	Serum	[87]

The first Luminex-based MDA, mPLEX-Flu, was developed to characterize the breadth and magnitude of the IgA, IgM and IgG antibodies against a large panel of whole HA proteins of multiple influenza virus types and subtypes in 2015 [67]. Our recent comprehensive studies, with novel statistical methods and a continuous readout across a 4.5 log range, indicated that MDA highly correlated with HAI and MN results, and with substantially better sensitivity and precision on account of continuous readout [64]. Furthermore, another study showed that using individual standard curves for each influenza HA strain in the mPlex-Flu assay to independently calculate IgG concentrations against each virus strain enables the direct comparison of serum anti-HA IgG concentrations against different influenza HA subtypes [76]. This ability addresses an essential issue for estimation and comparison of cross-reactivities of influenza antibody against multiple strains that has always plagued single-dimensional assays including HAI, MN and ELISA. The principle of the mPLEX-Flu assay is shown in Figure 1.

Based on the above characteristics of MDA, we use the example of a Luminex-based MDA, the mPLEX-Flu assay [67], to discuss the application of MDA on influenza vaccine development and basic viral immunity research. The major applications are summarized in Figure 2.

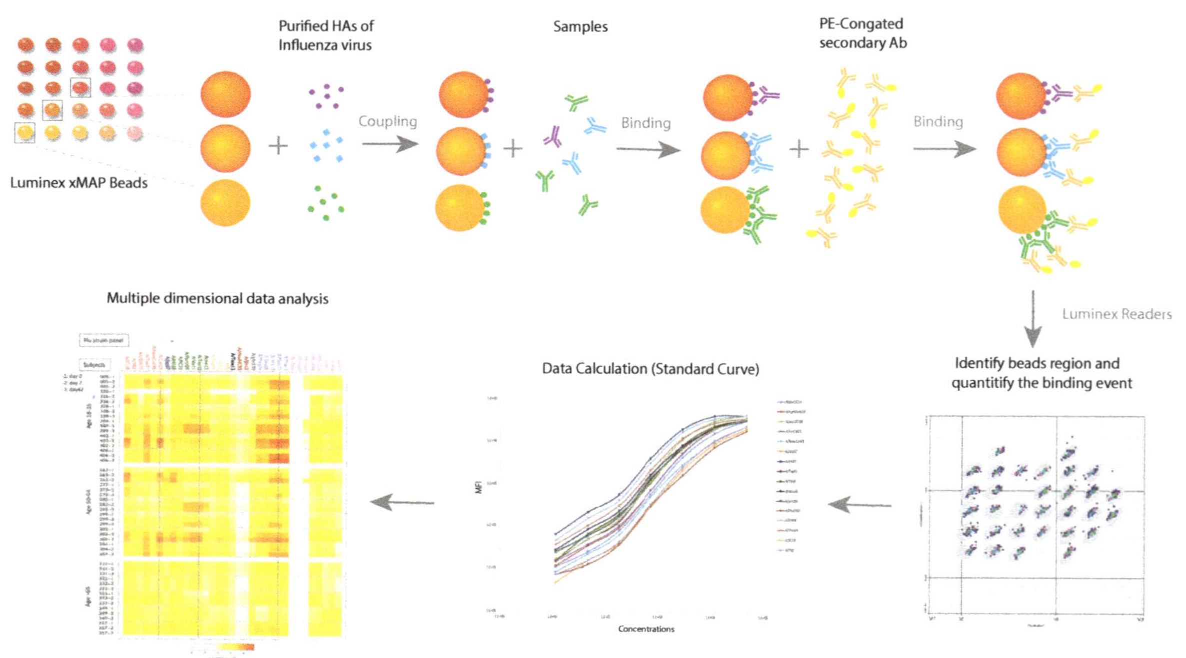

Figure 1. The principle of a multiple dimensional assay (MDA). The schematic diagram demonstrates the principle of the mPLEX-Flu assay, an example of a MDA, which is a Luminex based assay. Each color-coded xMAP bead is coupled with the purified recombinant hemagglutinin (rHAs) of one influenza virus strain. The different colors of influenza strain HA-specific beads are mixed and incubated with the serum sample. Bound anti-HA antibodies are subsequently detected using detection antibodies specific to each antibody isotope. The magnetic beads are read on a dual-laser flow-based Luminex reader. One laser classifies the bead and identifies the analyte being detected. The second laser determines the magnitude of the PE-derived signal, which is in direct proportion to the amount of analyte bound. Median fluorescence intensity (MFI) is converted to absolute concentration using a standard curve, at which point multiple dimensional data analysis can begin.

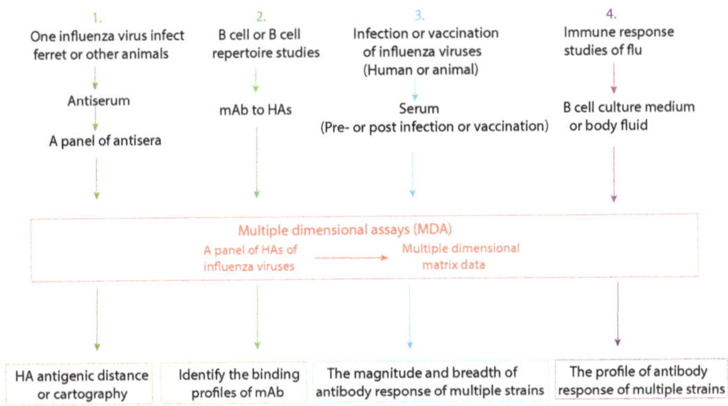

Figure 2. The current applications of multiple dimensional assays (MDAs). Four major influenza research applications of MDAs are listed, along with the multidimensional data set that each can generate.

5. Current Applications of MDA

5.1. Determination of the Antigenicity of HA of Influenza Virus

Antigenic cartography was first presented by Smith et al. in 2004 [88] as a way to quantify and visualize the antigenic differences in evolving flu strains. They used antigenic data from 35 years of H3 influenza surveillance, which consisted of hemagglutination inhibition (HI) titers multidimensional matrix data from 79 ferret polyclonal antisera against a panel of 273 viral isolates. They used multidimensional scaling analysis to adjust the position of viruses on an antigenic map such that the linear distance between two points reflects antigenic difference (calculated by comparing HI titers against other virus strains). The map revealed the high-level antigenic evolution of H3 influenza viruses from 1968 to 2002. To increase the efficiency and power of antigenic cartography, MDA could be used instead of HI to generate the multiple dimensional matrix data that could reveal the antigenic distance of HAs between the variants of influenza viruses, a technique we demonstrated in 2015 [67,71]. In the study of antigenic drift in 2015–2016 seasonal H1N1 viruses from the pandemic 2009 H1N1 virus [73], the mPLEX-Flu assay was sensitive enough to detect the antigenic difference of circulation isolates with H1 vaccine strain.

5.2. Identify the Binding Profiles of Broad Cross-Reactive mAb

Isolation and analysis of human monoclonal antibodies from the B cell repertoire of infected or vaccinated individuals is an important method to measure the B cell response against influenza virus [26,39,89,90]. As discussed above, broad cross-reactive antibodies (bcAbs) that are able to protect against influenza virus infection include the broad neutralizing antibodies (bnAbs) and other cross-reactive antibodies. bcAbs can be estimated by ELISA and MDA binding assays. Based on breadth of binding, bcAbs can be grouped as homosubtypic (cross-strain reactive within the same subtype groups), heterosubtypic (cross-reactive between subtypes), and heterophylogenic (cross-reactive across the phylogenic groups). MDAs can generate comprehensive high throughput data to determine the broad binding profile for mAb with tremendous efficiency. That will help accelerate research in this field. For example, KPF1, a human monoclonal Ab, was isolated by Kobies's lab from a subject who was immunized with the seasonal influenza inactivated vaccine [31]. Using the mPlex-Flu assay, the broad binding profile of KPF1 was efficiently clarified. Multiple dimensional data characterizing KPF1 clearly showed the magnitude and breadth of cross-reactivity of KPF1, and permitted visualization of results by a heat-map graph [31]. Importantly, the mplex-Flu assay revealed that the distinct binding profile of each mAb was different from the others, even though they were isolated from same influenza infection. This type of experiment demonstrates the utility of MDA assays in rapidly defining the immune repertoire landscape against multi-antigen HA proteins of multiple influenza strains.

5.3. Detection of the Magnitude and Breadth of Serologic Responses to Influenza Infection or Vaccination

The major goal in developing the mPlex-Flu assay is to quantitatively evaluate the cross-reactivity of influenza virus antibodies, including IgG, IgA and IgM isotypes. After we established and verified the mPlex-Flu assay, we applied it for detection of a breadth of cross-reactive Abs elicited by infection of influenza virus or vaccination with recombinant HA proteins [63] in mice and ferrets. The assay also provides a comprehensive and efficient way to evaluate the change of broad cross-reactive humoral immunity after influenza virus infection or vaccination in human clinical studies [72,91]. One of the most extensive benefits of the application of mPlex-Flu assay to studies of the antibody response of influenza is to provide more comprehensive data for baseline, before vaccination or infection [64,76]. The high throughput data of antibody titers helps to improve our understanding of the effects of influenza virus exposure history, or OAS that we discuss above, as essential factors that shape an individual's response to influenza vaccines or infections.

5.4. Detection of Antibodies in B Cell Culture Medium and Body Fluid

The high sensitivity and minimal sample size requirement enable MDA to quantitatively detect multiple influenza virus antibodies in samples other than serum (i.e., B-cell culture medium [72,74] and breast milk [75]), which contain low antibody concentration, and with small amount of sample volume, limit of detection for HAI, and MN single-panel traditional assays. Development of MBCs and activation of preexisting MBCs are essential features of the B cell response to influenza virus infection and vaccination [90,92]. Analyzing Abs in the culture supernatants of stimulated MBCs provides an alternative to ELISpot assay as a readout for HA-specific MBC responses, and facilitates a more comprehensive analysis of MBC repertoire [72]. HA-specific IgG concentrations in B cell culture medium are highly correlated with the frequencies of antigen-specific IgG secreting B cells derived from stimulated MBCs or plasmablasts [63,72,93].

As an example of the utility of combining MDA and in vitro culture experiments, we have previously analyzed low volume B cell culture samples using the mPLEX-Flu assay for changes in the size and character of HA-reactive MBC populations after H3N2 influenza infection [72] and seasonal flu vaccination [74] in a far more efficient and extensive way than could be accomplished with HAI or MN. We found that the H3-reactive IgG MBC population was expanded after infection induced reactivity to HA head and stalk domains, and head-reactive MBC populations were broad and reflected prior imprinting patterns of IgG production, which suggested that early-life H3N2 exposure affected H3 stalk-specific MBC expansion [72,74]. Similarly, a study examining the correlation between maternal and infant serum and maternal breast milk anti-influenza HA IgG and IgA patterns during the first 12 months of life showed that breast milk influenza HA-specific IgG and IgA antibody levels and patterns in breast milk were correlated with those in serum, except some H5, H4 and H9 HA head-specific Abs. A steady decay of infant influenza specific IgG levels by 6–8 months of age was also detected. This study strongly suggested that this new method could be used in a larger clinical study to understand the impact of maternal imprinting and temporary passive immunity on influenza immunity in the future [75].

6. Future Applications of MDA

6.1. Population Studies with Micro-Sampling Techniques

When coupled with low volume sampling methods, MDA has the potential to vastly increase subject sampling numbers for population based studies, while simultaneously yielding comprehensive data regarding IgG reactivity against multiple influenza strains. For example, a 2014 study used a protein microarray to monitor the trends of the 2009 influenza A (H1N1) pdm virus in 13 countries from five continents by screening bloodspots [78]. Similarly, a new technique called volumetric absorptive microsampling (VAMS), which provides for accurate sampling of a fixed blood volume (10 or 20 μL) on a volumetric swab and allows for long-term sample storage, has been used for peripheral blood sampling [94]. Combining this method with the mPlex-Flu assay enabled us to measure multidimensional anti-influenza IgG activity in whole blood samples collected by a finger-stick [95]. This study indicated that results from testing VAMS and traditional serum samples were highly correlated, both within subjects and across all influenza strains [95]. In addition, after adjustment for the hematocrit effects on the serum volume of whole blood samples, this new method accurately estimated the HA-specific IgG absolute concentration equivalent to that obtained with serum sampling methods. This novel approach provides a simple, accurate, low-cost tool for monitoring multidimensional anti-influenza hemagglutinin IgG responses in large population studies and clinical trials to comprehensively understand the effects of existing influenza virus antibodies on the immune response and new universal vaccine design.

The mPLEX-Flu assay can also be used to monitor the development of HA-specific antibodies against the influenza virus in infants. Tracking developing infant immunity to not only infection and vaccination strains but also other revolutionary strains would allow us to learn more about how OAS-type responses are first established. It would provide essential data for universal vaccine study and for understanding the mechanism of OAS.

6.2. Comprehensive Antigenic Study of HA Proteins

Right now, as discussed above, the broad neutralizing antibody (bnAb) activities still are the most important profiles of monoclonal antibodies to be considered. However, after the FluA-20 antibody was isolated, the broad cross-reactive antibodies (bcAbs) showed the protection against influenza virus that would likely be disregarded by traditional assays [51]. By contrast, MDA can detect non-neutralization activities of broad binding antibodies in a high throughput way. Absolutely, MDA will be a powerful serological assay for generating multidimensional data to exhibit the magnitude and breadth of binding to HAs from small amounts of sera.

6.3. Detecting Cross-Reactive Antibodies against Other Viral Proteins of Influenza Virus

Besides HA, neuraminidase (NA) also is an important target for inducing protective antibody-mediated responses [20]. Similarly, influenza virus M2 protein has an extracellular domain (M2e), which is highly conserved among influenza A viruses and B viruses. M2 is also being explored as a target for developing a "universal" vaccine to elicit the cross-protection against influenza. Unfortunately, little is known about the protective activity and broad cross-reaction of antibodies against these surface proteins.

At the same time, other internal proteins, such as nucleoprotein (NP) and the matrix protein 1 (M1), which are highly conserved between human seasonal and zoonotic influenza viruses [96], induce T-cell responses. These T-cell responses are shown to highly protect mice from the virus infection [97] and elicit robust CD8+ T cell responses across all human influenza A viruses [98]. After influenza virus infection, high circulating titers of NP Abs remain, and M1 antibodies can also be detected [96]. Currently, the effects of these antibodies against internal proteins on the T-cell response are unclear, especially the impact of pre-existing internal protein antibodies on the sequential humoral and cellular response.

Applying an expanded MDA panel that includes NA, NP and M1 protein-coupled beads in future influenza population surveys and clinical infection studies will allow us to evaluate antibodies against all these highly conserved proteins and HAs simultaneously. It would provide highly comprehensive data to help us to understand the T and B cell response to influenza virus infection, and also be beneficial for developing a "universal" influenza vaccine. While this review focuses on influenza, the underlying principles of MDA analysis apply equally to the study of immunity against other viruses that have multiple, antigenically similar strains.

7. Limitations of MDAs

As a novel technique, mPLEX-Flu and other MDAs also face some challenges and limitations. First, because MDAs detect the binding between the antibodies and HAs of influenza viruses, they are not able to directly reflect antibody protective function. By contrast, MN assays can estimate titers of neutralizing antibodies that inhibit influenza virus infection. However, our studies showed that data generated from mPLEX-Flu assay are highly correlated to HAI and MN data [31,63,64,76]. On the other side, HAI and MN assays have the chance to omit bcAbs that offer influenza virus protection through ADCC and ADCP, as discussed above. MDAs assay can be used to perform large scale screening, which can then be combined with other assays to identify and clarify the characters of antibodies.

Other limitations of MDAs are caused by the HA antigen coupled on the Luminex beads or microchips. First, most HA proteins used in the MDAs are expressed and purified by an insect cell baculovirus system that has to be glycosylated during post-translational modifications in infected insect cells. However, the insect cell glycosylation pathway is far simpler than those of human cells [99], even when both occur on the same N-glycosylation sites. Moreover, it has been reported that glycosylation of HAs is a crucial factor that needs to be considered when studying influenza infection and antigenic mutation (see review [100]). At present, it is not clear if differences in antigenic binding could be caused by the differences between insect cell-resourced HA proteins and those from human cells. Second, the density of HA proteins on the bead surface can vary, even when the coupling protocol and protein concentrations are kept consistent. This leads to error when comparing the binding between different influenza virus strains, which is similar to the error when comparing the HAI and MN titers between two viruses. To minimize this error, we established a unique method for generating standard curves for IgG binding to each virus strain [76].

8. Summary

The constantly changing HA antigenicity of influenza virus, along with the complexity of serological responses induced by the viruses in the human immune system, muddies efforts to interpret serology testing results. It had been commonly accepted that assessing the antibody response against only vaccine strain viral HAs is too restricted for understanding this complexity. Understanding the effects of pre-existing antibodies and cross-reactive antibodies against multiple strains' HAs is becoming increasingly enticing in the influenza B cell immunity research field. MDAs and systems serology, the novel technologies combined with multidimensional data, computer modeling, and bioinformatics, are groundbreaking new tools for influenza vaccine study. They will open a novel comprehensive view to investigate the B cell response to influenza virus and be a powerful tool for universal vaccine development.

Funding: This work was supported by the National Institutes of Health Institute of Allergy, Immunology and Infectious Diseases grant including R01 AI129518-01, AI098112, and R21 AI138500.

Acknowledgments: We would like to thank Jonathan Kasper for discussing and careful reading of this manuscript.

Conflicts of Interest: All authors declare no conflict of interest.

Abbreviations

The following abbreviations are used in this manuscript:

MDA	Multiple dimensional assay
mAb	monoclonal antibody
HA	hemagglutinin
NA	neuraminidase
MBC	memory B cell
HAI	hemagglutinin inhibition assay
MN	microneutralization assay
bcAb	broad cross-reactive antibody
bnAb	broad neutralizing antibody

References

1. Iuliano, A.D.; Roguski, K.M.; Chang, H.H.; Muscatello, D.J.; Palekar, R.; Tempia, S.; Cohen, C.; Gran, J.M.; Schanzer, D.; Cowling, B.J.; et al. Estimates of global seasonal influenza-associated respiratory mortality: A modelling study. *Lancet* **2018**, *391*, 1285–1300. [CrossRef]
2. Koel, B.F.; Burke, D.F.; Bestebroer, T.M.; Van Der Vliet, S.; Zondag, G.C.M.; Vervaet, G.; Skepner, E.; Lewis, N.S.; Spronken, M.I.J.; Russell, C.A.; et al. Substitutions Near the Receptor Binding Site Determine Major Antigenic Change During Influenza Virus Evolution. *Science* **2013**, *342*, 976–979. [CrossRef] [PubMed]
3. Hilleman, M.R. Realities and enigmas of human viral influenza: Pathogenesis, epidemiology and control. *Vaccine* **2002**, *20*, 3068–3087. [CrossRef]
4. Carrat, F.; Flahault, A. Influenza vaccine: The challenge of antigenic drift. *Vaccine* **2007**, *25*, 6852–6862. [CrossRef]
5. Lambert, L.C.; Fauci, A.S. Influenza Vaccines for the Future. *New Engl. J. Med.* **2010**, *363*, 2036–2044. [CrossRef] [PubMed]
6. CDC (Centers for Disease Control and Prevention). *Selecting Viruses for the Seasonal Influenza Vaccine*; CDC: Atlanta, GA, USA, 2018.
7. Erbelding, E.J.; Post, D.J.; Stemmy, E.J.; Roberts, P.C.; Augustine, A.D.; Ferguson, S.; Paules, C.I.; Graham, B.S.; Fauci, A.S. A Universal Influenza Vaccine: The Strategic Plan for the National Institute of Allergy and Infectious Diseases. *J. Infect. Dis.* **2018**, *218*, 347–354. [CrossRef]
8. Zost, S.J.; Parkhouse, K.; Gumina, M.E.; Kim, K.; Diaz Perez, S.; Wilson, P.C.; Treanor, J.J.; Sant, A.J.; Cobey, S.; Hensley, S.E. Contemporary H3N2 influenza viruses have a glycosylation site that alters binding of antibodies elicited by egg-adapted vaccine strains. *Proc. Natl. Acad. Sci. USA* **2017**, *114*, 12578–12583. [CrossRef]
9. Francis, T., Jr. On the Doctrine of Original Antigenic Sin. *Proc. Am. Philos. Soc.* **1960**, *104*, 572–578. [CrossRef]
10. Gostic, K.M.; Ambrose, M.; Worobey, M.; Lloyd-Smith, J.O. Potent protection against H5N1 and H7N9 influenza via childhood hemagglutinin imprinting. *Science* **2016**, *354*, 722–726. [CrossRef]
11. Schulman, J.L.; Kilbourne, E.D. Induction of Partial Specific Heterotypic Immunity in Mice by a Single Infection with Influenza a Virus. *J. Bacteriol.* **1965**, *89*, 170–174.
12. Nguyen, H.H.; Zemlin, M.; Ivanov, I.; Andrasi, J.; Zemlin, C.; Vu, H.L.; Schelonka, R.; Schroeder, H.W., Jr.; Mestecky, J. Heterosubtypic immunity to influenza A virus infection requires a properly diversified antibody repertoire. *J. Virol.* **2007**, *81*, 9331–9338. [CrossRef] [PubMed]
13. Hirst, G.K. The quantitative determination of influenza virus and antibodies by means of red cell agglutination. *J. Exp. Med.* **1942**, *75*, 49–64. [CrossRef] [PubMed]
14. Hirst, G.K. Adsorption of influenza hemagglutinins and virus by red blood cells. *J. Exp. Med.* **1942**, *76*, 195–209. [CrossRef] [PubMed]
15. Frank, A.L.; Puck, J.; Hughes, B.J.; Cate, T.R. Microneutralization test for influenza A and B and parainfluenza 1 and 2 viruses that uses continuous cell lines and fresh serum enhancement. *J. Clin. Microbiol.* **1980**, *12*, 426–432.

16. Remarque, E.J.; de Bruijn, I.A.; Boersma, W.J.; Masurel, N.; Ligthart, G.J. Altered antibody response to influenza H1N1 vaccine in healthy elderly people as determined by HI, ELISA, and neutralization assay. *J. Med. Virol.* **1998**, *55*, 82–87. [CrossRef]
17. Kilbourne, E.D. Taxonomy and Comparative Virology of the Influenza Viruses. In *Influenza*; Springer: Boston, MA, USA, 1987; pp. 25–32.
18. Nachbagauer, R.; Choi, A.; Hirsh, A.; Margine, I.; Iida, S.; Barrera, A.; Ferres, M.; Albrecht, R.A.; García-Sastre, A.; Bouvier, N.M. Defining the antibody cross-reactome directed against the influenza virus surface glycoproteins. *Nat. Immunol.* **2017**, *18*, 464. [CrossRef]
19. Yamashita, M.; Krystal, M.; Fitch, W.M.; Palese, P. Influenza B virus evolution: Co-circulating lineages and comparison of evolutionary pattern with those of influenza A and C viruses. *Virology* **1988**, *163*, 112–122. [CrossRef]
20. Krammer, F.; Fouchier, R.A.M.; Eichelberger, M.C.; Webby, R.J.; Shaw-Saliba, K.; Wan, H.; Wilson, P.C.; Compans, R.W.; Skountzou, I.; Monto, A.S. NAction! How Can Neuraminidase-Based Immunity Contribute to Better Influenza Virus Vaccines? *MBio* **2018**, *9*. [CrossRef]
21. Mair, C.M.; Ludwig, K.; Herrmann, A.; Sieben, C. Receptor binding and pH stability—How influenza A virus hemagglutinin affects host-specific virus infection. *Biochim. Biophys. Acta* **2014**, *1838*, 1153–1168. [CrossRef]
22. Brandenburg, B.; Koudstaal, W.; Goudsmit, J.; Klaren, V.; Tang, C.; Bujny, M.V.; Korse, H.J.; Kwaks, T.; Otterstrom, J.J.; Juraszek, J. Mechanisms of hemagglutinin targeted influenza virus neutralization. *PLoS ONE* **2013**, *8*, e80034. [CrossRef]
23. Coudeville, L.; Bailleux, F.; Riche, B.; Megas, F.; Andre, P.; Ecochard, R. Relationship between haemagglutination-inhibiting antibody titres and clinical protection against influenza: Development and application of a bayesian random-effects model. *BMC Med. Res. Methodol.* **2010**, *10*, 18. [CrossRef]
24. Caton, A.J.; Brownlee, G.G.; Yewdell, J.W.; Gerhard, W. The antigenic structure of the influenza virus A/PR/8/34 hemagglutinin (H1 subtype). *Cell* **1982**, *31*, 417–4127. [CrossRef]
25. Wiley, D.C.; Wilson, I.A.; Skehel, J.J. Structural identification of the antibody-binding sites of Hong Kong influenza haemagglutinin and their involvement in antigenic variation. *Nature* **1981**, *289*, 373–378. [CrossRef] [PubMed]
26. Angeletti, D.; Gibbs, J.S.; Angel, M.; Kosik, I.; Hickman, H.D.; Frank, G.M.; Das, S.R.; Wheatley, A.K.; Prabhakaran, M.; Leggat, D.J.; et al. Defining B cell immunodominance to viruses. *Nat. Immunol.* **2017**, *18*, 456–463. [CrossRef] [PubMed]
27. Xu, R.; Wilson, I.A. Structural characterization of an early fusion intermediate of influenza virus hemagglutinin. *J. Virol.* **2011**, *85*, 5172–5182. [CrossRef]
28. Yassine, H.M.; Boyington, J.C.; McTamney, P.M.; Wei, C.J.; Kanekiyo, M.; Kong, W.P.; Gallagher, J.R.; Wang, L.; Zhang, Y.; Joyce, M.G.; et al. Hemagglutinin-stem nanoparticles generate heterosubtypic influenza protection. *Nat. Med.* **2015**, *21*, 1065–1070. [CrossRef]
29. Krammer, F.; Palese, P. Universal Influenza Virus Vaccines That Target the Conserved Hemagglutinin Stalk and Conserved Sites in the Head Domain. *J. Infect. Dis.* **2019**, *219*, S62–S67. [CrossRef]
30. Ellebedy, A.H. Immunizing the Immune: Can We Overcome Influenza's Most Formidable Challenge? *Vaccines* **2018**, *6*, 68. [CrossRef]
31. Nogales, A.; Piepenbrink, M.S.; Wang, J.; Ortega, S.; Basu, M.; Fucile, C.F.; Treanor, J.J.; Rosenberg, A.F.; Zand, M.S.; Keefer, M.C.; et al. A Highly Potent and Broadly Neutralizing H1 Influenza-Specific Human Monoclonal Antibody. *Sci. Rep.* **2018**, *8*, 4374. [CrossRef]
32. Hong, M.; Lee, P.S.; Hoffman, R.M.B.; Zhu, X.; Krause, J.C.; Laursen, N.S.; Yoon, S.I.; Song, L.; Tussey, L.; Crowe, J.E.; et al. Antibody Recognition of the Pandemic H1N1 Influenza Virus Hemagglutinin Receptor Binding Site. *J. Virol.* **2013**, *87*, 12471–12480. [CrossRef]
33. Whittle, J.R.R.; Zhang, R.; Khurana, S.; King, L.R.; Manischewitz, J.; Golding, H.; Dormitzer, P.R.; Haynes, B.F.; Walter, E.B.; Moody, M.A.; et al. Broadly neutralizing human antibody that recognizes the receptor-binding pocket of influenza virus hemagglutinin. *Proc. Natl. Acad. Sci. USA* **2011**, *108*, 14216–14221. [CrossRef] [PubMed]

34. Xu, R.; Krause, J.C.; McBride, R.; Paulson, J.C.; Crowe, J.E.; Wilson, I.A. A recurring motif for antibody recognition of the receptor-binding site of influenza hemagglutinin. *Nat. Struct. Mol. Biol.* **2013**. *20*, 363–370. [CrossRef]
35. Ekiert, D.C.; Kashyap, A.K.; Steel, J.; Rubrum, A.; Bhabha, G.; Khayat, R.; Lee, J.H.; Dillon, M.A.; O'Neil, R.E.; Faynboym, A.M.; et al. Cross-neutralization of influenza A viruses mediated by a single antibody loop. *Nature* **2012**, *489*, 526–532. [CrossRef] [PubMed]
36. Lee, P.S.; Yoshida, R.; Ekiert, D.C.; Sakai, N.; Suzuki, Y.; Takada, A.; Wilson, I.A. Heterosubtypic antibody recognition of the influenza virus hemagglutinin receptor binding site enhanced by avidity. *Proc. Natl. Acad. Sci. USA* **2012**, *109*, 17040–17045. [CrossRef] [PubMed]
37. Lee, P.S.; Ohshima, N.; Stanfield, R.L.; Yu, W.; Iba, Y.; Okuno, Y.; Kurosawa, Y.; Wilson, I.A. Receptor mimicry by antibody F045–092 facilitates universal binding to the H3 subtype of influenza virus. *Nat. Commun.* **2014**, *5*. [CrossRef]
38. Yoshida, R.; Igarashi, M.; Ozaki, H.; Kishida, N.; Tomabechi, D.; Kida, H.; Ito, K.; Takada, A. Cross-Protective Potential of a Novel Monoclonal Antibody Directed against Antigenic Site B of the Hemagglutinin of Influenza A Viruses. *PLoS Pathog.* **2009**, *5*, e1000350. [CrossRef]
39. Lee, J.; Boutz, D.R.; Chromikova, V.; Joyce, M.G.; Vollmers, C.; Leung, K.; Horton, A.P.; DeKosky, B.J.; Lee, C.H.; Lavinder, J.J.; et al. Molecular-level analysis of the serum antibody repertoire in young adults before and after seasonal influenza vaccination. *Nat. Med.* **2016**, *22*, 1456–1464. [CrossRef]
40. Ekiert, D.C.; Bhabha, G.; Elsliger, M.A.; Friesen, R.H.; Jongeneelen, M.; Throsby, M.; Goudsmit, J.; Wilson, I.A. Antibody recognition of a highly conserved influenza virus epitope. *Science* **2009**, *324*, 246–251. [CrossRef]
41. Sui, J.; Hwang, W.C.; Perez, S.; Wei, G.; Aird, D.; Chen, L.M.; Santelli, E.; Stec, B.; Cadwell, G.; Ali, M.; et al. Structural and functional bases for broad-spectrum neutralization of avian and human influenza A viruses. *Nat. Struct. Mol. Biol.* **2009**, *16*, 265–273. [CrossRef]
42. Kashyap, A.K.; Steel, J.; Rubrum, A.; Estelles, A.; Briante, R.; Ilyushina, N.A.; Xu, L.; Swale, R.E.; Faynboym, A.M.; Foreman, P.K.; et al. Protection from the 2009 H1N1 Pandemic Influenza by an Antibody from Combinatorial Survivor-Based Libraries. *PLoS Pathog.* **2010**, *6*, e1000990. [CrossRef]
43. Ekiert, D.C.; Friesen, R.H.; Bhabha, G.; Kwaks, T.; Jongeneelen, M.; Yu, W.; Ophorst, C.; Cox, F.; Korse, H.J.; Brandenburg, B.; et al. A highly conserved neutralizing epitope on Group 2 influenza A viruses. *Science* **2011**, *333*, 843–850. [CrossRef] [PubMed]
44. Friesen, R.H.E.; Lee, P.S.; Stoop, E.J.M.; Hoffman, R.M.B.; Ekiert, D.C.; Bhabha, G.; Yu, W.; Juraszek, J.; Koudstaal, W.; Jongeneelen, M.; et al. A common solution to Group 2 influenza virus neutralization. *Proc. Natl. Acad. Sci. USA* **2014**, *111*, 445–450. [CrossRef] [PubMed]
45. Kallewaard, N.L.; Corti, D.; Collins, P.J.; Neu, U.; McAuliffe, J.M.; Benjamin, E.; Wachter-Rosati, L.; Palmer-Hill, F.J.; Yuan, A.Q.; Walker, P.A.; et al. Structure and Function Analysis of an Antibody Recognizing All Influenza A Subtypes. *Cell* **2016**, *166*, 596–608. [CrossRef] [PubMed]
46. Lang, S.; Xie, J.; Zhu, X.; Wu, N.C.; Lerner, R.A.; Wilson, I.A. Antibody 27F3 Broadly Targets Influenza A Group 1 and 2 Hemagglutinins through a Further Variation in V H 1-69 Antibody Orientation on the HA Stem. *Cell Rep.* **2017**, *20*, 2935–2943. [CrossRef]
47. Corti, D.; Voss, J.; Gamblin, S.J.; Codoni, G.; Macagno, A.; Jarrossay, D.; Vachieri, S.G.; Pinna, D.; Minola, A.; Vanzetta, F.; et al. A Neutralizing Antibody Selected from Plasma Cells That Binds to Group 1 and Group 2 Influenza A Hemagglutinins. *Science* **2011**, *333*, 850–856. [CrossRef]
48. Dreyfus, C.; Laursen, N.S.; Kwaks, T.; Zuijdgeest, D.; Khayat, R.; Ekiert, D.C.; Lee, J.H.; Metlagel, Z.; Bujny, M.V.; Jongeneelen, M.; et al. Highly Conserved Protective Epitopes on Influenza B Viruses. *Science* **2012**, *337*, 1343–1348. [CrossRef]
49. Henry Dunand, C.J.; Leon, P.E.; Huang, M.; Choi, A.; Chromikova, V.; Ho, I.Y.; Tan, G.S.; Cruz, J.; Hirsh, A.; Zheng, N.Y.; et al. Both Neutralizing and Non-Neutralizing Human H7N9 Influenza Vaccine-Induced Monoclonal Antibodies Confer Protection. *Cell Host Microbe* **2016**, *19*, 800–813. [CrossRef]
50. Mullarkey, C.E.; Bailey, M.J.; Golubeva, D.A.; Tan, G.S.; Nachbagauer, R.; He, W.; Novakowski, K.E.; Bowdish, D.M.; Miller, M.S.; Palese, P.; et al. Broadly Neutralizing Hemagglutinin Stalk-Specific Antibodies Induce Potent Phagocytosis of Immune Complexes by Neutrophils in an Fc-Dependent Manner. *MBio* **2016**, *7*. [CrossRef]

51. Bangaru, S.; Lang, S.; Schotsaert, M.; Vanderven, H.A.; Zhu, X.; Kose, N.; Bombardi, R.; Finn, J.A.; Kent, S.J.; Gilchuk, P.; et al. A Site of Vulnerability on the Influenza Virus Hemagglutinin Head Domain Trimer Interface. *Cell* **2019**, *177*, 1136–1152.e18. [CrossRef]
52. Xu, R.; Ekiert, D.C.; Krause, J.C.; Hai, R.; Crowe, J.E.; Wilson, I.A. Structural basis of preexisting immunity to the 2009 H1N1 pandemic influenza virus. *Science* **2010**, *328*, 357–360. [CrossRef]
53. Dawood, F.S.; Iuliano, A.D.; Reed, C.; Meltzer, M.I.; Shay, D.K.; Cheng, P.Y.; Bandaranayake, D.; Breiman, R.F.; Brooks, W.A.; Buchy, P.; et al. Estimated global mortality associated with the first 12 months of 2009 pandemic influenza A H1N1 virus circulation: A modelling study. *Lancet Infect. Dis.* **2012**, *12*, 687–695. [CrossRef]
54. Nguyen, A.M.; Noymer, A. Influenza mortality in the United States, 2009 pandemic: Burden, timing and age distribution. *PLoS ONE* **2013**, *8*, e64198. [CrossRef] [PubMed]
55. De St.Groth, S.F.; Webster, R. Disquisitions on original antigenic sin: I. Evidence in man. *J. Exp. Med.* **1966**, *124*, 331–345. [CrossRef] [PubMed]
56. Zarnitsyna, V.I.; Ellebedy, A.H.; Davis, C.; Jacob, J.; Ahmed, R.; Antia, R. Masking of antigenic epitopes by antibodies shapes the humoral immune response to influenza. *Philos. Trans. R. Soc. Lond. B Biol. Sci.* **2015**, *370*. [CrossRef] [PubMed]
57. Miller, M.S.; Gardner, T.J.; Krammer, F.; Aguado, L.C.; Tortorella, D.; Basler, C.F.; Palese, P. Neutralizing antibodies against previously encountered influenza virus strains increase over time: A longitudinal analysis. *Sci. Transl. Med.* **2013**, *5*, 198ra107. [CrossRef]
58. Zhang, A.; Stacey, H.D.; Mullarkey, C.E.; Miller, M.S. Original Antigenic Sin: How First Exposure Shapes Lifelong Anti-Influenza Virus Immune Responses. *J. Immunol.* **2019**, *202*, 335–340. [CrossRef]
59. Lessler, J.; Riley, S.; Read, J.M.; Wang, S.; Zhu, H.; Smith, G.J.D.; Guan, Y.; Jiang, C.Q.; Cummings, D.A.T. Evidence for Antigenic Seniority in Influenza A (H3N2) Antibody Responses in Southern China. *PLoS Pathog.* **2012**, *8*, e1002802. [CrossRef]
60. Huijskens, E.G.; Reimerink, J.; Mulder, P.G.; van Beek, J.; Meijer, A.; de Bruin, E.; Friesema, I.; de Jong, M.D.; Rimmelzwaan, G.F.; Peeters, M.F.; et al. Profiling of humoral response to influenza A(H1N1)pdm09 infection and vaccination measured by a protein microarray in persons with and without history of seasonal vaccination. *PLoS ONE* **2013**, *8*, e54890. [CrossRef]
61. Koopmans, M.; de Bruin, E.; Godeke, G.J.; Friesema, I.; van Gageldonk, R.; Schipper, M.; Meijer, A.; van Binnendijk, R.; Rimmelzwaan, G.F.; de Jong, M.D.; et al. Profiling of humoral immune responses to influenza viruses by using protein microarray. *Clin. Microbiol. Infect.* **2012**, *18*, 797–807. [CrossRef]
62. Te Beest, D.; de Bruin, E.; Imholz, S.; Wallinga, J.; Teunis, P.; Koopmans, M.; van Boven, M. Discrimination of influenza infection (A/2009 H1N1) from prior exposure by antibody protein microarray analysis. *PLoS ONE* **2014**, *9*, e113021. [CrossRef]
63. Wang, J.; Hilchey, S.P.; DeDiego, M.; Perry, S.; Hyrien, O.; Nogales, A.; Garigen, J.; Amanat, F.; Huertas, N.; Krammer, F.; et al. Broad cross-reactive IgG responses elicited by adjuvanted vaccination with recombinant influenza hemagglutinin (rHA) in ferrets and mice. *PLoS ONE* **2018**, *13*, e0193680. [CrossRef] [PubMed]
64. Li, D.; Wang, J.; Garigen, J.; Treanor, J.J.; Zand, M.S. Continuous Readout versus Titer-Based Assays of Influenza Vaccine Trials: Sensitivity, Specificity, and False Discovery Rates. *Comput. Math. Methods Med.* **2019**, *2019*, 9287120. [CrossRef] [PubMed]
65. Legutki, J.B.; Magee, D.M.; Stafford, P.; Johnston, S.A. A general method for characterization of humoral immunity induced by a vaccine or infection. *Vaccine* **2010**, *28*, 4529–4537. [CrossRef] [PubMed]
66. Nakajima, R.; Supnet, M.; Jasinskas, A.; Jain, A.; Taghavian, O.; Obiero, J.; Milton, D.K.; Chen, W.H.; Grantham, M.; Webby, R.; et al. Protein Microarray Analysis of the Specificity and Cross-Reactivity of Influenza Virus Hemagglutinin-Specific Antibodies. *mSphere* **2018**, *3*. [CrossRef] [PubMed]
67. Wang, J.; Hilchey, S.P.; Hyrien, O.; Huertas, N.; Perry, S.; Ramanunninair, M.; Bucher, D.; Zand, M.S. Multi-Dimensional Measurement of Antibody-Mediated Heterosubtypic Immunity to Influenza. *PLoS ONE* **2015**, *10*, e0129858. [CrossRef]
68. Luminex Instruments. Available online: https://www.luminexcorp.com/instruments/ (accessed on 5 May 2019).

69. Watson, D.S.; Reddy, S.M.; Brahmakshatriya, V.; Lupiani, B. A multiplexed immunoassay for detection of antibodies against avian influenza virus. *J. Immunol. Methods* **2009**, *340*, 123–131. [CrossRef]
70. Keynan, Y.; Bodnarchuk, T.; Wayne, S.; Li, Y.; Fowke, K.R. Evaluation of influenza-specific humoral response by microbead array analysis. *Can. J. Infect. Dis. Med. Microbiol.* **2011**, *22*, 25–29. [CrossRef]
71. Zand, M.S.; Wang, J.; Hilchey, S. Graphical Representation of Proximity Measures for Multidimensional Data: Classical and Metric Multidimensional Scaling. *Math. J.* **2015**, *17*. [CrossRef]
72. Tesini, B.L.; Kanagaiah, P.; Wang, J.; Hahn, M.; Halliley, J.L.; Chaves, F.A.; Nguyen, P.Q.T.; Nogales, A.; DeDiego, M.L.; Anderson, C.S.; et al. Broad Hemagglutinin-Specific Memory B Cell Expansion by Seasonal Influenza Virus Infection Reflects Early-Life Imprinting and Adaptation to the Infecting Virus. *J. Virol.* **2019**, *93*. [CrossRef]
73. Clark, A.M.; DeDiego, M.L.; Anderson, C.S.; Wang, J.; Yang, H.; Nogales, A.; Martinez-Sobrido, L.; Zand, M.S.; Sangster, M.Y.; Topham, D.J. Antigenicity of the 2015-2016 seasonal H1N1 human influenza virus HA and NA proteins. *PLoS ONE* **2017**, *12*, e0188267. [CrossRef]
74. Huang, J.; Hilchey, S.P.; Wang, J.; Gerigan, J.; Zand, M.S. IL-15 enhances cross-reactive antibody recall responses to seasonal H3 influenza viruses in vitro. *F1000Research* **2017**, *6*, 2015. [CrossRef] [PubMed]
75. Järvinen, K.M.; Wang, J.; Seppo, A.E.; Zand, M. Novel multiplex assay for profiling influenza antibodies in breast milk and serum of mother-infant pairs. *F1000Research* **2019**. [CrossRef]
76. Li, D.; Wang, J.; Treanor, J.J.; Zand, M.S. Improved Specificity and False Discovery Rates for Multiplex Analysis of Changes in Strain-Specific Anti-Influenza IgG. *Comput. Math. Methods Med.* **2019**, *2019*, 3053869. [CrossRef] [PubMed]
77. Germeraad, E.; Achterberg, R.; Venema, S.; Post, J.; de Leeuw, O.; Koch, G.; van der Wal, F.J.; Beerens, N. The development of a multiplex serological assay for avian influenza based on Luminex technology. *Methods* **2019**, *158*, 54–60. [CrossRef] [PubMed]
78. De Bruin, E.; Loeber, J.G.; Meijer, A.; Castillo, G.M.; Cepeda, M.L.G.; Torres-Sepúlveda, M.R.; Borrajo, G.J.C.; Caggana, M.; Giguere, Y.; Meyer, M.; et al. Evolution of an influenza pandemic in 13 countries from 5 continents monitored by protein microarray from neonatal screening bloodspots. *J. Clin. Virol.* **2014**, *61*, 74–80. [CrossRef] [PubMed]
79. Freidl, G.S.; de Bruin, E.; van Beek, J.; Reimerink, J.; de Wit, S.; Koch, G.; Vervelde, L.; van den Ham, H.J.; Koopmans, M.P. Getting more out of less—A quantitative serological screening tool for simultaneous detection of multiple influenza A hemagglutinin-types in chickens. *PLoS ONE* **2014**, *9*, e108043. [CrossRef] [PubMed]
80. Freidl, G.S.; van den Ham, H.J.; Boni, M.F.; de Bruin, E.; Koopmans, M.P. Changes in heterosubtypic antibody responses during the first year of the 2009 A(H1N1) influenza pandemic. *Sci. Rep.* **2016**, *6*, 20385. [CrossRef]
81. Te Beest, D.E.; de Bruin, E.; Imholz, S.; Koopmans, M.; van Boven, M. Heterosubtypic cross-reactivity of HA1 antibodies to influenza A, with emphasis on nonhuman subtypes (H5N1, H7N7, H9N2). *PLoS ONE* **2017**, *12*, e0181093. [CrossRef]
82. Desbien, A.L.; Van Hoeven, N.; Reed, S.J.; Casey, A.C.; Laurance, J.D.; Baldwin, S.L.; Duthie, M.S.; Reed, S.G.; Carter, D. Development of a high density hemagglutinin protein microarray to determine the breadth of influenza antibody responses. *Biotechniques* **2013**, *54*, 345–348. [CrossRef]
83. Price, J.V.; Jarrell, J.A.; Furman, D.; Kattah, N.H.; Newell, E.; Dekker, C.L.; Davis, M.M.; Utz, P.J. Characterization of influenza vaccine immunogenicity using influenza antigen microarrays. *PLoS ONE* **2013**, *8*, e64555. [CrossRef]
84. Meade, P.; Latorre-Margalef, N.; Stallknecht, D.E.; Krammer, F. Development of an influenza virus protein microarray to measure the humoral response to influenza virus infection in mallards. *Emerg. Microbes Infect.* **2017**, *6*, e110. [CrossRef] [PubMed]
85. Freidl, G.S.; Binger, T.; Muller, M.A.; de Bruin, E.; van Beek, J.; Corman, V.M.; Rasche, A.; Drexler, J.F.; Sylverken, A.; Oppong, S.K.; et al. Serological evidence of influenza A viruses in frugivorous bats from Africa. *PLoS ONE* **2015**, *10*, e0127035. [CrossRef] [PubMed]
86. Mace, C.R.; Topham, D.J.; Mosmann, T.R.; Quataert, S.A.; Treanor, J.J.; Miller, B.L. Label-free, arrayed sensing of immune response to influenza antigens. *Talanta* **2011**, *83*, 1000–1005. [CrossRef] [PubMed]
87. Bucukovski, J.; Latorre-Margalef, N.; Stallknecht, D.E.; Miller, B.L. A Multiplex Label-Free Approach to Avian Influenza Surveillance and Serology. *PLoS ONE* **2015**, *10*, e0134484. [CrossRef]

88. Smith, D.J. Mapping the Antigenic and Genetic Evolution of Influenza Virus. *Science* **2004**, *305*, 371–376. [CrossRef]
89. Bangaru, S.; Zhang, H.; Gilchuk, I.M.; Voss, T.G.; Irving, R.P.; Gilchuk, P.; Matta, P.; Zhu, X.; Lang, S.; Nieusma, T.; et al. A multifunctional human monoclonal neutralizing antibody that targets a unique conserved epitope on influenza HA. *Nat. Commun.* **2018**, *9*. [CrossRef]
90. Wrammert, J.; Smith, K.; Miller, J.; Langley, W.A.; Kokko, K.; Larsen, C.; Zheng, N.Y.; Mays, I.; Garman, L.; Helms, C.; et al. Rapid cloning of high-affinity human monoclonal antibodies against influenza virus. *Nature* **2008**, *453*, 667–671. [CrossRef]
91. Hill, D.L.; Pierson, W.; Bolland, D.J.; Mkindi, C.; Carr, E.J.; Wang, J.; Houard, S.; Wingett, S.W.; Audran, R.; Wallin, E.F.; et al. The adjuvant GLA-SE promotes human Tfh cell expansion and emergence of public TCRbeta clonotypes. *J. Exp. Med.* **2019**. [CrossRef]
92. Pinna, D.; Corti, D.; Jarrossay, D.; Sallusto, F.; Lanzavecchia, A. Clonal dissection of the human memory B-cell repertoire following infection and vaccination. *Eur. J. Immunol.* **2009**, *39*, 1260–1270. [CrossRef]
93. Sangster, M.Y.; Baer, J.; Santiago, F.W.; Fitzgerald, T.; Ilyushina, N.A.; Sundararajan, A.; Henn, A.D.; Krammer, F.; Yang, H.; Luke, C.J.; et al. B cell response and hemagglutinin stalk-reactive antibody production in different age cohorts following 2009 H1N1 influenza virus vaccination. *Clin. Vaccine Immunol.* **2013**, *20*, 867–876. [CrossRef]
94. Kok, M.G.M.; Fillet, M. Volumetric absorptive microsampling: Current advances and applications. *J. Pharm. Biomed. Anal.* **2018**, *147*, 288–296. [CrossRef]
95. Wang, J.; Li, D.; Wiltse, A.; Emo, J.; Hilchey, S.P.; Zand, M.S. Application of Volumetric Absorptive Micro Sampling to Measure Multidimensional Anti-Influenza Hemagglutinin Igg Antibodies by MPlex-Flu Assay. *J. Clin. Transl. Sci.* **2019**, *3*, 332–343. [CrossRef]
96. Sebastian, S.; Lambe, T. Clinical Advances in Viral-Vectored Influenza Vaccines. *Vaccines* **2018**, *6*, 29. [CrossRef] [PubMed]
97. Kreijtz, J.H.; Bodewes, R.; van Amerongen, G.; Kuiken, T.; Fouchier, R.A.; Osterhaus, A.D.; Rimmelzwaan, G.F. Primary influenza A virus infection induces cross-protective immunity against a lethal infection with a heterosubtypic virus strain in mice. *Vaccine* **2007**, *25*, 612–620. [CrossRef] [PubMed]
98. Quinones-Parra, S.; Grant, E.; Loh, L.; Nguyen, T.H.; Campbell, K.A.; Tong, S.Y.; Miller, A.; Doherty, P.C.; Vijaykrishna, D.; Rossjohn, J.; et al. Preexisting CD8+ T-cell immunity to the H7N9 influenza A virus varies across ethnicities. *Proc. Natl. Acad. Sci. USA* **2014**, *111*, 1049–1054. [CrossRef] [PubMed]
99. Shi, X.; Jarvis, D.L. Protein N-glycosylation in the baculovirus-insect cell system. *Curr. Drug Targets* **2007**, *8*, 1116–1125. [CrossRef] [PubMed]
100. York, I.A.; Stevens, J.; Alymova, I.V. Influenza virus N-linked glycosylation and innate immunity. *Biosci. Rep.* **2019**, *39*. [CrossRef]

pathogens

MDPI

Review

Host Single Nucleotide Polymorphisms Modulating Influenza A Virus Disease in Humans

Aitor Nogales [1,*] and Marta L. DeDiego [2,*]

1 Center for Animal Health Research, INIA-CISA, 28130 Madrid, Spain
2 Department of Molecular and Cell Biology, Centro Nacional de Biotecnología (CNB-CSIC), Campus Universidad Autónoma de Madrid, 28049 Madrid, Spain
* Correspondence: nogales.aitor@inia.es (A.N.); Marta.Lopez@cnb.csic.es (M.L.D.)

Received: 11 September 2019; Accepted: 28 September 2019; Published: 30 September 2019

Abstract: A large number of human genes associated with viral infections contain single nucleotide polymorphisms (SNPs), which represent a genetic variation caused by the change of a single nucleotide in the DNA sequence. SNPs are located in coding or non-coding genomic regions and can affect gene expression or protein function by different mechanisms. Furthermore, they have been linked to multiple human diseases, highlighting their medical relevance. Therefore, the identification and analysis of this kind of polymorphisms in the human genome has gained high importance in the research community, and an increasing number of studies have been published during the last years. As a consequence of this exhaustive exploration, an association between the presence of some specific SNPs and the susceptibility or severity of many infectious diseases in some risk population groups has been found. In this review, we discuss the relevance of SNPs that are important to understand the pathology derived from influenza A virus (IAV) infections in humans and the susceptibility of some individuals to suffer more severe symptoms. We also discuss the importance of SNPs for IAV vaccine effectiveness.

Keywords: Influenza A virus (IAV); innate immunity; single nucleotide polymorphisms (SNPs); Influenza vaccine

1. Introduction

1.1. *Influenza A Virus (IAV)*

Influenza A viruses (IAV) belong to the *Orthomyxoviridae* family, and they contain a single-stranded (ss) negative-sense viral (v)RNA genome formed by eight segments that are encapsidated into particles with an envelope (Figure 1A). Each of the vRNA segments contains a long central coding region flanked at 5′ and 3′ termini by non-coding regions (NCRs), which work as promoters to initiate viral RNA synthesis (transcription and replication). Moreover, the packaging signals playing a role in the efficient encapsidation of the viral segments into nascent virions, are located at the 3′ and 5′ end of the coding regions (Figure 1B) [1]. Structurally, vRNAs form viral ribonucleoprotein complexes (vRNPs), where vRNAs are coated with multiple subunits of the viral nucleoprotein (NP) and are associated with the heterotrimeric polymerase, which contains the polymerase basic 2 and 1 (PB2 and PB1, respectively) and acidic (PA) proteins (Figure 1A) [2–4]. Each vRNP acts as an independent transcription-replication unit using an uncommon mechanism among negative-sense RNA viruses, given that viral RNA synthesis occurs in the infected-cells nucleus. vRNAs are used as templates by the viral polymerase to synthesize two positive-sense RNA molecules, the complementary RNAs (cRNAs), from which the same viral polymerase synthesizes more copies of genomic vRNA, and the mRNAs for viral protein synthesis [1–6]. The small IAV genome encodes for up to 16 viral proteins through multiple mechanisms, although there are some differences between strains. For that genome plasticity,

IAV take advantage of multiple strategies, such as alternative splicing, frameshift mechanisms, and overlapping open reading frames (ORFs) [7–9]. In addition, the coding capability of the viral genome is extended by encoding multifunctional proteins that act at different steps during virus infection.

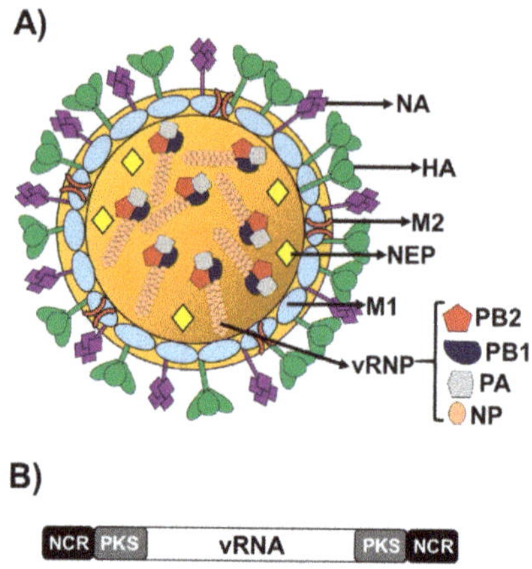

Figure 1. Influenza A virus structure and genome organization. (**A**) Virion structure: Influenza A viruses include a lipid envelope containing the two viral glycoproteins (HA, green and NA, purple). The ion channel M2 (red) protein is also located in the membrane. Under the viral bilayer is located a protein layer composed of the M1 (light blue) protein, and the NEP (yellow). Inside the virion are located the eight vRNA segments coated by the NP (pink) as viral vRNP complexes and associated with the viral polymerase complex made of the three polymerase subunits PB2 (red), PB1 (blue) and PA (gray). Viral components in the vRNP and in the viral particle are indicated. (**B**) Genome organization: Influenza A virus contains eight ss, negative-sense, viral RNA segments (PB2, PB1, PA, HA, NP, NA, M, and NS). Each viral segment contains non-coding regions (NCR, black) and the packaging signals (PKS, gray) at the 3′ and 5′ termini in each of the viral segments.

The viral envelope is decorated with the two viral glycoproteins hemagglutinin (HA) and neuraminidase (NA) at a ratio of approximately four to one, respectively [10,11]. HA envelope protein mediates virus entry by binding to sialic acid-containing cell receptors, and then fusing endosomal and viral membranes during endocytosis [12,13], while NA is required for viral release from infected host cells, and it acts as a receptor destroying enzyme, cleaving terminal sialic acid residues from glycoproteins present at the cell surface [14–16]. The matrix 2 (M2) protein is also found in the viral membrane, although in much lower abundance than HA or NA glycoproteins. M2 is a small transmembrane protein that forms a proton-selective ion channel in the viral envelope. M2 promotes uncoating of the vRNPs after membrane fusion and the protein has also an essential role in viral assembly and release [17]. Under the viral envelop, there is an inner shell composed of the matrix 1 (M1) protein, which interacts in the virion with the vRNP and the HA and NA proteins. M1 apart from being a membrane-associated scaffold factor of the virion, acts as a crucial factor for different viral processes during infection, including virion assembly and budding [18–20]. The nonstructural (NS) gene or segment 8 of IAV encodes an mRNA transcript that is alternatively spliced to express two viral proteins, the nonstructural protein 1 (NS1), produced from a continuous primary transcript, and the nuclear export protein (NEP), which is produced by an alternatively processed transcript, using a weak 5′ splice site. NEP is also located in the virion and may interact with M1 in the viral particle [21–23] (Figure 1A). During the infection, NEP is responsible for the nuclear export of synthetized vRNP, ensuring that the vRNPs are available for packaging [24]. Moreover, NEP has also other functions during IAV infection, contributing to viral budding and to regulate viral RNA synthesis. NS1 is a multifunctional protein and a key viral factor that counteracts the host antiviral responses. NS1 has been shown to inhibit the production of interferon (IFN), the activity and expression of multiple interferon-induced genes (ISG) and the processing and nuclear transport of host mRNAs causing cellular shut-off [25,26]. Segment 3 of IAV also encodes two proteins, the polymerase component PA and PA-X. PA is translated directly from the PA mRNA, whereas PA-X is translated using a +1 frameshift mechanism from the same open reading frame (ORF) [9]. Synergistically with NS1, PA-X is also able to block the cellular antiviral responses by inhibiting host protein expression. Moreover, the PA-X protein has been shown to modulate host inflammation, immune responses, apoptosis, and virus pathogenesis [25–30].

1.2. Influenza Virus Importance in Human Health

Human IAV infections cause contagious respiratory diseases associated with mild to severe respiratory illness or even death, and they are considered as an important public health threat worldwide, which also results in significant economic losses [31–33]. IAV are divided into multiple subtypes, based on the HA and NA glycoproteins. Currently, there are 18 HA (H1 to H18) and 11 NA (N1 to N11), but the growing IAV surveillance programs and sequencing technologies could increase the number of subtypes in the following years. IAV can infect a wide range of avian and mammalian species, although the natural reservoirs of IAV are shorebirds and wild waterfowls [34–37]. Among all the HA and NA subtypes, only H3N2 and H1N1 IAV subtypes are circulating in human beings and they are responsible for annual recurrent epidemics that affect the entire world [38,39]. Seasonal influenza infections are prevented and controlled through annual vaccination campaigns to decrease IAV infections and viral transmission as well as to reduce their negative impact in the global economy. However, although vaccination remains the most effective approach to protect the population from seasonal infections, the effectiveness of current vaccination approaches is suboptimal [16,31–33,39–44]. Thus, the production of improved prophylactic approaches, including universal vaccines, are highly desired. Concerns associated with IAV are further aggravated by the adaptive capacity of the viruses to infect new hosts or escape to the immune system, as well as their ability to transmit efficiently in the population and the limited therapeutic options to treat viral infections [14,16,25,45].

Because of the ability of IAV to modify their genome using two main evolutionary mechanisms, antigenic drift and shift, viruses encoding novel antigenic proteins to which the population has limited or no preexisting immunity can be generated [10,31,37,40]. For that reason, seasonal vaccines have to be reformulated yearly to guarantee that the viral glycoproteins (HA and NA) in the vaccine match seasonal viruses circulating worldwide [38,43,46]. In addition, IAV variability can lead to the generation of new virus strains with pandemic potential. For example, the first IAV pandemic of this century occurred in 2009 and it is estimated that in approximately one year, the pandemic 2009 H1N1 (pH1N1) IAV infected more than 600,000 human beings, causing near 16,000 deaths in over 200 countries [40,41]. In addition, although only H1N1 and H3N2 are circulating in humans, the avian H5, H7, and H9 subtypes eventually cross the species barrier to infect humans, representing a new and serious public health problem [13,37,47–49].

1.3. Innate Immunity in IAV Infections

The cellular defense mechanisms provided by the innate immune system are a formidable barrier to inhibit virus infections [50] and involve the recognition of pathogen-associated molecular patterns (PAMPs) by pattern recognition receptors (PRRs). This recognition leads to the activation of signaling pathways and the production and secretion of IFNs of type I (IFNα and IFNβ) and III (IFNλ2 or IL-28A, IFNλ3 or IL-28B, and IFNλ1 or IL-29), and chemokines and cytokines involved in inflammatory processes [50]. IAV RNAs are mainly recognized by the endosomal, membrane-associated PRR Toll-like receptors (TLRs) 3 (double-stranded RNAs, dsRNAs) or 7/8 (ssRNAs), respectively [50,51], by the cytoplasmic PRR retinoic acid-inducible gene I (RIG-I), which detects dsRNA and 5′- triphosphates of the negative ssRNA viral genome [50,52], generated during replication of multiple viruses, by the NOD-like receptor family member NOD-, LRR- and pyrin domain-containing 3 (NLRP3), which recognizes various stimuli (see below) [53] and by the absent in melanoma 2 (AIM2) protein, recognizing not well-characterized influenza stimuli [54]. The result of PRR detection of viral PAMPs is the activation of multiple transcription factors, such as the nuclear factor kappa β (NF-κB), the activator protein 1 (AP-1), and IFN regulatory factors (IRF)-3 and IRF-7, which are responsible for the transcription of IFNs [50,55,56] and pro-inflammatory cytokines [57].

Secreted type I and III IFNs signal through different receptors in a paracrine or autocrine way to induce the transcription of IFN-stimulated genes (ISGs), several of which counteract viral replication [50,56,58]. Just as an example mentioned below, IFITM3 is an ISG playing antiviral roles against influenza virus infection and other viruses [59]. Type I and III IFNs signaling pathways lead to the post-translational phosphorylation of the signal transducer and activator of transcription (STAT) 1 and 2 transcription factors [60], being the tyrosine kinase 2 (TYK2) and Janus protein tyrosine kinase 1 (JAK1) critical for the phosphorylation [61]. Moreover, STAT1 is phosphorylated by IKKε during IFN signaling and this step is important for the IFN-inducible innate immune response [62,63]. Upon phosphorylation, STAT1 and STAT2 associate with IRF-9 forming the heterotrimeric ISG factor 3 (ISGF3) complex [60]. This heterotrimeric complex then translocates to the nucleus, and binds to IFN-stimulated response elements (ISREs) located in the promoters of ISGs, up-regulating their expression [60,64].

Inflammatory cytokines, such as interleukins (IL)-1A IL-1B and tumor necrosis factor (TNF)-α contribute to the proliferation and migration of different immune cells, such as monocytes, macrophages, neutrophils, and natural killer (NK) cells, to the infected tissue. NK cells have the ability to kill virus-infected cells, are important for the activation of a protective cytotoxic T lymphocyte (CTL) response [65], and NK-cell IFN-γ production is augmented by T-cell IL-2 production in recall responses [66]. Neutrophils and resident alveolar macrophages are also important for virus clearance, due to their ability to destroy infected cells [67]. In addition, cytokine signaling improves dendritic cells (DC) maturation, increasing the induction of adaptive immune responses by antigen presentation and co-stimulation [68,69]. These adaptive immune responses initiated upon innate immune activation are required for protection and viral clearance [70].

NLRP3 is expressed by myeloid cells such as macrophages, monocytes, neutrophils, and dendritic cells [71] or by human bronchial epithelial cells [72]. Upon stimulation, NLRP3 activates the inflammasome system, activating caspase-1 and leading to pro-inflammatory processes through the processing and activation of proIL-1B, proIL-18, and proIL-33 [73]. NLRP3 senses IAV dsRNA [74], and PB1-F2 protein [75]. Furthermore, protein flux through the viral M2 ion channel activity in the trans-Golgi network activates NLRP3, leading to inflammasome activation [76]. In addition to NLRP3 activation, IAV activates the inflammasomes through AIM2, increasing IAV-induced lung injury and mortality [54].

The complement system is an important branch of innate immunity that plays an essential role in the clearance of pathogens. The complement system is triggered by three main pathways, the classical, the lectin, and the alternative pathways [77]. The first two pathways are activated with the help of pattern recognition molecules, whereas the alternative pathway is activated spontaneously. Interestingly, it is known that viruses are recognized by the three pathways. In the classical pathway, the C1 complex recognizes antigen-antibody complexes, which are formed on the pathogen surface. C1QBP (Complement C1q Binding Protein) can bind to the globular heads of C1q molecules, activating the classical pathway [78]. On the other hand, in the lectin pathway, the mannan-binding lectin (MBL)/ficolin/mannan-binding lectin serin protease (MAP) complex recognizes specific carbohydrates on the pathogen surface. Complexes activated after the classical and lectin pathways, cleave C4 and C2, resulting in the generation of C4bC2a (C3 convertase). In the alternative pathway, spontaneous hydrolysis of native C3 results in the formation of C3b-like C3 that binds factor B and after cleavage by factor D forms the initial C3 convertase [77]. The three pathways converge at the cleavage of C3 into C3a and C3b by C3 convertases (C4b, 2a and C3b, Bb). Then, the C3b molecules formed bind covalently to the C3-convertases forming the C5-convertases that cleave C5 into C5a and C5b. CD55 blocks C3 and C5 activation by preventing the formation of new C3 and C5 convertases [79]. C5b starts the formation of C5b-9 or the membrane attack complex (MAC). Next, C8 binds to the membrane attached trimer and begins binding and polymerization of C9 that is inserted into the membrane, inducing virolysis [77].

Unregulated complement activation could play a central role in the acute lung injury (ALI) pathology induced by highly pathogenic viruses, including severe acute respiratory syndrome (SARS) coronavirus and avian IAV H5N1, and H7N9 [80]. In virus-induced acute lung diseases, high levels of chemotactic, and anaphylatoxic C5a can be generated as a result of excessive complement triggering and causing a "cytokine storm". Accordingly, the blockade of C5a signaling has been involved in treating the ALI induced by highly pathogenic viruses [80].

1.4. Single Nucleotide Polymorphisms (SNPs)

Currently, particular attention is being paid to single nucleotide polymorphisms (SNPs) that are loci within the genome of an organism in which two or more alleles can exist. SNPs affect a single nucleotide or base pair and they are one of the most frequent types of genetic variations in the genome [81–83]. SNPs need to be presented into the population with a frequency equal to or greater than 1% to be considered as polymorphisms. There are multiple types of SNPs, depending on their location that can be in different regions of the genes such as promoters, exons, introns or UTRs (Figure 2). SNPs in coding regions are classified as synonymous, when a nucleotide substitution does not change the amino acid sequence of the encoded protein, although other effects, such as changes in mRNA structure or folding may account for variation in protein expression. On the other hand, non-synonymous SNPs are divided in missense or nonsense. In the first case, nucleotide substitution results in the change of one amino acid for another, affecting the protein sequence coded by a gene and therefore may lead to its dysfunction. In contrast, nonsense mutations are produced when instead of substituting one amino acid for another, the altered gene contains an early stop codon in the ORF or a stop codon is abrogated, producing an elongated protein. This type of mutations results in shortened or elongated proteins leading typically to nonfunctional proteins. The functional role of SNPs in coding areas of the genome can be easily analyzed by studying the gene products. However, most SNPs fall within non-coding genome regions, therefore, predicting their effects is challenging. For example, SNPs in the promoter regions could affect their activity and regulation producing changes in gene expression levels. SNPs in UTRs or intron regions have been related with an effect in protein translation or the production of splice variants of transcripts, leading to longer or shorter protein sequences, respectively.

In summary, SNPs may influence gene regulation, the structure and stability of RNA, the expression of RNAs or proteins, the conformation and function of proteins, etc. Thus, the identification of SNPs in genes and the analysis of their effects may lead us to better understand gene function or their impact on human health [84]. In fact, SNPs that are or could be important for multiple human pathologies, such as cancer, diabetes, heart disease, schizophrenia, blood-pressure homeostasis, and autoimmune or metabolic diseases, have been identified [85–91]. Moreover, some described SNPs increase the human susceptibility to getting infected by viruses, bacteria or other pathogens [84,86,92–97]. Advanced sequencing and bioinformatics technologies have allowed the identification of a large number of human SNPs whose information is accessible in the databases. Nevertheless, the biological significance and function for most of the SNPs found in the human genome remain unknown. Currently, the scientific community recognizes the importance of this kind of genome variations that can act as biological markers and assist researchers in multiple aspects, such as: (1) Locate genes associated with multiple diseases, (2) anticipate an individual's response to a specific infection, (3) predict population responses to several treatments such as drugs or vaccines, (4) design individualized therapies, (5) identify markers for medical testing, (6) perform pharmacogenetic studies, etc. This review focuses on the role of known SNPs on IAV infection, as well as their impact on the effectiveness of vaccines against IAV.

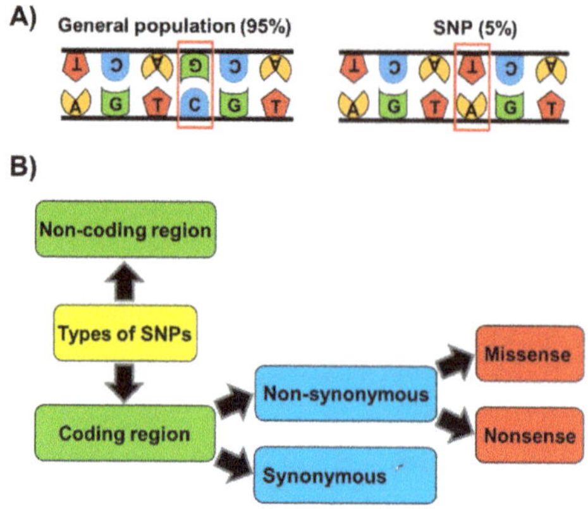

Figure 2. Single nucleotide polymorphisms (SNPs). (**A**) An SNP is a variation on a single nucleotide which may occur at some specific point in the genome and that causes variations in DNA sequences between members of the same species. (**B**) Types of SNPs: DNA variation can be located in non-coding or coding regions. SNPs within a coding sequence can be synonymous if they do not produce an amino acid change (silent mutation), or non-synonymous if they affect the protein sequence. Non-synonymous changes can be divided into missense (producing an amino acid change in the protein) or nonsense (producing a truncated or longer protein).

2. SNPs in Host Genes Affecting IAV Disease

Risk factors, including underlying co-morbidities, age, and pregnancy, affect IAV susceptibility, but do not explain all the conditions under which serious IAV-associated disease can occur, making likely that SNPs in viral and host genes affect IAV susceptibility and the outcome of the disease. In fact, there are some examples of the presence of SNPs in host genes affecting influenza severity (Table 1), which will be discussed in this review. SNPs affecting IAV disease have been found in genes recognizing viral components, in transcription factors important for IFN production and signaling, in ISGs with antiviral activities, and in genes involved in inflammation.

TLR3 recognizes dsRNA, one of the IAV replication intermediate products, and in turn activates IFN production, leading to an antiviral response. A missense mutation (F303S) of the TLR3 gene was found in one out of three patients developing IAV-associated encephalopathy (IAE), a neurological consequence of severe viral infection [98]. Assays in tissue culture cells showed that a TLR3 receptor encoding the missense F303S mutation was impaired in activating the transcription factor NF-κB, and in triggering downstream signaling via the IFNβ receptor, indicating that this genetic polymorphism could lead to increased IAV replication [98]. In a study of 51 Italian children diagnosed with IAV H1N1 infection, an additional TLR3 SNP (rs5743313, genotype C/T) was identified [99]. This TLR3 SNP was found in all the children developing IAV-associated pneumonia (18 cases). However, the SNP was found in significantly less proportion in children with milder disease, suggesting a link between TLR3 and IAV pathogenicity. Furthermore, in a multicenter study involving 275 adult cases of avian H7N9 and pH1N1 IAV, in mainland China and Hong Kong, the TLR3 CC rs5743313 SNP was associated with fatal cases [100].

In addition to IAV, there are other examples of SNPs in TLR3 or TLR3 signaling genes affecting viral infections. For instance, susceptibility to Chikungunya virus (CHIKV) infection is highly increased in human and mouse cells with defective TLR3 molecules [101]. Furthermore, *TLR3* SNPs, rs3775292, and rs6552950, leading to unknown functional consequences, were associated with an increased risk of CHIKV disease occurrence [101]. Patients with impaired TLR3-mediated responses show an elevated susceptibility to Herpes Simplex-1 Virus (HSV-1)-mediated encephalitis by encoding TLR3-deficient alleles [102,103], or by encoding deficient TRAF3, TBK1 and TRIF molecules, leading to impaired TLR-3 signaling [104–106]. In a Saudi Arabian population, the TLR3 rs78726532 SNP was strongly associated with Hepatitis B (HBV) and Hepatitis C (HCV) virus infections when compared to that in healthy control subjects [107,108]. The TLR3 rs5743314 C allele was also associated with HCV-related liver disease progression (cirrhosis and hepatocellular carcinoma) [107]. However, the functional effects of these SNPs seem to be unknown.

RIG-I detects dsRNA and 5′-triphosphates of the negative ssRNA IAV genome, leading to innate immune responses activation [52]. A Caucasian male patient with severe IAV H1N1 infection during the 2009 swine flu pandemic showed two heterozygous variants (one in each chromosome): p.R71H (SNP rs72710678) and p.P885S (SNP rs138425677), located, respectively, in the caspase activation and recruitment domain (CARD) and RNA binding domains of RIG-I [109]. These variants significantly decreased the recognition function of RIG-I, and therefore, patient cells proved impaired antiviral responses to RIG-I ligands and elevated proinflammatory responses to IAV, providing evidence for dysregulation of the innate immune response and increased immunopathology [109]. These results suggest that these RIG-I polymorphisms may have contributed to severe IAV outcome in this patient and reinforce that RIG-I variants should be evaluated in future studies of host factors affecting ssRNA virus infections.

IRF-7 is a transcription factor that increases interferon (IFN) production in response to viruses [110–112]. A patient suffering from an unusual life-threatening disease after pH1N1 infection encodes homozygous null mutations in the IRF-7 factor. Both IRF-7 alleles from this patient encode mutations c.1228T>G/T (F410V) and c.1261C>T/C (Q421X), which are mutations decreasing the ability of IRF-7 to induce the transcription of IFN genes after IAV infections [113]. These findings suggest that IRF-7-dependent production of type I and III IFNs is required for controlling IAV infections in humans. The rare allele A of two IRF-7 SNPs, rs12272434 and rs12290989, both located at exon/intron boundaries, were significantly associated with impaired levels of IFNα production by human plasmacytoid dendritic cells (pDCs) in response to human immunodeficiency virus 1 (HIV-1) infection [114]. Therefore, these polymorphisms may affect the ability of human subjects to control HIV-1 infections, reinforcing the role of IRF-7 in controlling viral infections. However, the effect of these SNPs should be further studied.

IRF-9 is a transcription factor essential for IFN signaling and the transcriptional induction of ISGs [60]. STAT1 and STAT2, when phosphorylated, associate with IRF-9 to form a heterotrimeric ISG factor 3 (ISGF3) complex [60], which translocates to the nucleus, and binds ISREs present in the promoters of ISGs, up-regulating their transcription [60,64]. A homozygous, loss-of-function mutation in IRF-9 was described in a child born to first-cousin Algerian parents and living in France affected by a severe pulmonary influenza infection [115]. In particular, the homozygous mutation (c.991G>A) occurred in the final nucleotide of exon 7 and disrupted the essential splice site at the boundary of exon 7 and intron 7, leading to deleted IRF-9 proteins. The consequence of this mutation was an impaired activation of IRF-9, and therefore, an impaired transcription of ISGs, many of which show antiviral activities [115]. Similarly, a family in which several members showed a surprising susceptibility to infection by different viruses, including IAV, also showed to be IRF9 deficient [116]. The index patient, a boy with 10 years born at term from healthy consanguineous parents (first cousins of Portuguese origin and residing in Venezuela) encoded a homozygous splicing mutation in the IRF9 gene. The mutation, c.577+1G>T, was located in the donor splice site of introns 5 and 6, leading to transcripts lacking exon 5. IRF9 protein expression was undetectable in cells transfected with the c.577+1G>T IRF9 construct, suggesting that either the protein was quickly degraded or the mRNA was not translated. Again, IRF9-deficient cells showed a profound defect in inducing the expression of multiple ISGs [116]. Collectively, these findings show that human IRF9- and ISGF3-dependent type I and III IFN responsive pathways are essential for controlling viral infections, including IAV.

The antiviral protein IFITM3 is an ISG which abrogates the release of IAV content from late endosomes into the cytoplasm [59]. In addition, IFITM3 promotes the survival of mouse lung-resident CD8+ T cells following IAV challenge, which may help clear the infection [117]. Furthermore, mice in which the expression of IFITM3 is abolished, showed severe disease after IAV infection, compared to parental mice [118]. One of the clearest associations of SNPs in genes affecting influenza severity is located in the ISG IFITM3. The human IFITM3 gene is encoded by two exons and is predicted to encode two splice variants that differ in the first amino-terminal 21 amino acids. Different studies have described the effect of IFITM3 SNPs in influenza disease severity. Northern European patients infected with IAV pH1N1 2009 virus requiring hospitalization showed over-representation of the SNP rs12252 in the IFITM3 gene, in which the majority T allele is replaced for a minority C allele [118]. This leads to an alteration of the first splice acceptor site, originating an IFITM3 protein lacking the first 21 amino acids (NΔ21) due to the protein starting from an alternative start codon. According to these results suggesting that this SNP could affect influenza disease, the minority (CC) variant rendered homozygous cells more susceptible to IAV infection, and this susceptibility correlated with decreased levels of IFITM3 protein expression in comparison to the majority (TT) variant cells [118]. Furthermore, cells expressing the NΔ21 protein showed an impaired ability to restrict viral replication when compared to wild-type IFITM3 cells [118]. This data is consistent with previous results which show that the amino-terminal 21 amino acids of IFITM3 are relevant for attenuating vesicular stomatitis virus (VSV) replication in vitro [119]. Moreover, the CC genotype was found in 25% of Chinese patients showing mild disease after pH1N1 virus infection compared to 69% in patients developing a severe pH1N1 virus infection. In addition, the CC genotype was estimated to confer a six-fold increased risk for severe infection than the CT and TT genotypes [120], reinforcing the idea that IFITM3 is a factor affecting human IAV disease [121]. In another study, over-representation of the IFITM3 CC genotype was detected among fatal cases of Chinese patients infected with IAV pH1N1 and H7N9 viruses [100], and in a more general study, including twelve studies published before February 2018 with more than 16,000 subjects, revealed increased risk of severe influenza in both the East Asian and White populations in the subjects encoding the IFITM3 CC genotype [122].

Another important SNP (rs34481144) associated with risk of severe influenza in humans from the United States (US) infected with seasonal IAVs is located in the 5′-UTR of the IFITM3 gene [123,124]. This SNP affected IFITM3 expression being the risk allele associated with lower mRNA expression. The mechanism for this lower mRNA expression involves the decreased IRF-3 binding and increased binding of the transcriptional repressor CCCTC-binding factor (CTCF) in promoter-binding assays for the risk allele [123]. Moreover, the risk allele disrupted a CpG site that becomes differentially methylated in CD8+ T cell subsets, leading to less CD8+ T cells in the airways during natural influenza infection in the carriers of the risk allele, and suggesting that a critical role for IFITM3 may be to promote immune cell persistence at mucosal sites [123].

Interleukins 1A and 1B (IL-1A and IL-1B, respectively) are inflammatory cytokines that play critical roles in recruiting immune and inflammatory cells and developing adaptive immune responses. Furthermore, accumulating evidence suggests that both cytokines play central roles in innate immunity against viral infections [125]. The frequencies of SNP (allele C) located 31 base pairs upstream from the transcription start site (rs1143627), on the IL-1B promoter were associated with increased risk of influenza disease in Chinese subjects [126]. This nucleotide change is localized in a TATA-box motif of IL-1B and modulates the transcription activity of IL-1B by binding to multiple transcription factors [127]. The allele T of rs1143627 enhanced IL-1B protein expression, as indicated by several reports [128]. People carrying allele T showed a higher IL-1B expression, which could lead to increased IFNγ production, which promotes virus clearance [129]. In contrast, expression of IL-1B may be decreased in individuals who carry allele C, leading to a weaker immune response during viral infection. In addition, a T allele in IL-1A gene (SNP rs17561) increased the risk of IAV pH1N1 susceptibility, as observed in Chinese subjects [126]. The SNP rs17561 introduces a nonsynonymous mutation (A114S) in IL-1A protein, suggesting that this genetic variant may lead to a functional variation in host susceptibility to pH1N1. Nevertheless, the molecular mechanism needs to be evaluated and the real risk of these alleles should be analyzed in larger populations.

TNF-α is a pro-inflammatory cytokine which orchestrates the host´s defense. A minor allele (A) at position -238 of TNF (SNP rs361525) was more frequent in Greek patients infected with pH1N1 virus compared to control subjects [130], and developing pneumonia was more uncommon in Greek and Mexican subjects with no copies of the minor allele compared to subjects with at least one copy of the minor allele [130,131], leading to the hypothesis that this SNP allele could be linked with an elevated susceptibility to infection with the pH1N1 virus [124,130]. Decreased TNF-α expression was observed in subjects encoding the minor allele at position -238 [92]. This may explain how SNPs leading to lower production of TNF-α may predispose to more severe clinical symptoms following IAV infections. However, the TNF-α rs 1800629 minor A allele, associated with higher levels of TNF-α expression, was associated with susceptibility to Japanese Encephalitis virus infection in an Indian population [132]. The TNF-α rs 1800629 minor A allele was a risk factor to develop liver cirrhosis and hepatocellular carcinoma following HBV infection in a Han Chinese population [133], suggesting that the protective or deleterious roles of TNF-α expression may vary depending on the infecting virus.

Chemokine receptor 5 (CCR5) is expressed mainly on macrophages, T cells, and dendritic cells. CCR5 mediates leukocyte chemotaxis in response to its ligands, including MIP-1a, MIP-1b, and RANTES. It can help direct multiple immune cell subsets, including regulatory T cells or Th17 cells to sites of infection, supporting the antiviral immune response. Evidence in humans support that homozygosity for the CCR5-Δ32 allele, a naturally occurring polymorphism of CCR5 encoding a 32-bp deletion, prevents its expression on the cell surface, and is linked with an elevated susceptibility to West Nile virus (WNV) [134] and with increased severity of illness among patients infected with pH1N1 [135], although this evidence is modest due to the limited number of subjects analyzed. In contrast, homozygous carriers of the Δ32 mutation are resistant to HIV-1 infection because this molecule, absent in the cell surface in subjects encoding the deletion, is a molecule normally used by HIV-1 to enter CD4+ T cells [136].

CD55 is an important complement regulatory protein which blocks C3 and C5 activation by preventing the formation of new C3 and C5 convertases, two proteases involved in inflammation and complement activation. Consequently, CD55 protects cells from complement attack and decreases amplification of the complement cascade [79]. The CD55 SNP (rs2564978, genotype T/T) was significantly associated with severe IAV infection in Chinese patients infected with pH1N1 2009 virus [137] and was associated with increased death risk in Greek patients [138]. The rs2564978 SNP of CD55 is located in the minimal promoter region [139] and individuals with this genotype showed significantly lower levels of CD55 expression in comparison to those with the more frequent allele [137]. Therefore, patients who carry the T/T genotype may have more robust complement activation during IAV infection, resulting in enhanced inflammation and disease severity [47,79]. According to these results, the polymorphism rs2564978 in gene CD55 was linked to disease severity in adult Chinese cases of avian (H7N9) and human pH1N1 IAV in another study [100]. However, these findings need to be confirmed in bigger cohorts.

C1QBP can bind to the globular heads of C1q molecules, activating the classical pathway of complement [78]. An increased risk of severe disease after IAV infection was found in patients homozygous for the minor allele of the SNP rs3786054 in European and Mexican populations [138,140]. However, the effect of this SNPs on gene expression and function is undescribed.

Soluble pattern-recognition molecules, forming part of the innate immune system, can neutralize IAV infection. Particularly, the serum mannose-binding lectin (MBL), several secreted human C-type lectins of the collectin family, collectin 11, and the pulmonary surfactant proteins (SP) –A1, –A2, and –D (SFTPA1, SFTPA2, and SFTPD, respectively), may neutralize IAV infectivity in vitro [141]. Mice lacking SP-A or SP-D were more susceptible to IAV infection, indicating that SPs exert relevant roles against IAV infection [142–144]. Two frequent SP-A2 (SFTPA2) missense alleles (rs1965708-C, leading to the mutation Q223K and rs1059046-A, leading to the mutation T9N) were associated with acute respiratory failure, mechanical ventilation, and acute respiratory distress syndrome after infection with pH1N1 2009 virus in a Spanish population [145].

In addition to C-type lectins, S-type lectins have been described, such as galectins, which recognize galactose-containing oligosaccharides present in the cellular plasma membranes and in viruses, such as IAV. Importantly, intranasal treatment of galectin-1 enhanced survival of mice infected with IAV by reducing viral load, apoptosis, and inflammation in the lung [146]. Moreover, galectin-1 knockout mice showed increased susceptibility to influenza virus infection than wild-type mice [146]. To study human genetic susceptibility to avian IAV H7N9 infection, a genome-wide association study involving 106 heavily-exposed healthy poultry Chinese workers and 102 IAV H7N9 patients was performed [147]. Functional variants of galectin-1 gene, including rs4820294 and rs13057866, causing increased expression levels of galectin-1 expression, may confer more protection from IAV H7N9 infection to the carriers of these variants [147].

The cleavage of the IAV HA by host proteases is critical for viral infectivity. TMPRSS2 is a type II transmembrane serine protease family member, which was shown to activate HA proteins of multiple human IAVs in tissue culture cells. Furthermore, deletion of Tmprss2 in mice impairs the spread of H1N1 influenza viruses, including the pH1N1 2009 swine IAV [148]. In addition, bodyweight loss and survival after H3N2 IAV infection were less severe in Tmprss2 mutant mice compared to wild type mice [148]. The genetic predisposition to severe pH1N1 2009 influenza virus was evaluated in Chinese human subjects, finding that the GG genotype of rs2070788, leading to increased expression of TMPRSS2, was a risk variant to severe pH1N1 influenza [149]. Furthermore, rs2070788 and rs383510, both of them associated with increased gene expression, were significantly associated with the susceptibility to IAV H7N9 [149].

Table 1. Single nucleotide polymorphisms associated with susceptibility and severity of influenza infections.

Gene	Function	SNPs (Type)	References
TLR-3	Recognizes dsRNA, triggering IFN production.	rs not annotated; F303S (NonSyn). rs5743313 (NCR).	[98] [99,100]
RIG-I	Detects dsRNA and 5′- triphosphates of the negative ssRNA IAV genome, leading to innate immune responses activation.	rs72710678; R71H (NonSyn). rs138425677; P885S (NonSyn).	[109]
IRF-7	Transcription factor that increases IFN production in response to viruses.	rs786205223; F410V (NonSyn) rs375323253; Q421X (NonSyn)	[113]
IRF-9	Transcription factor essential for IFN signaling and the transcriptional induction of ISGs.	c.991G>A occurred in the final nucleotide of exon 7 and disrupted the essential splice site at the boundary of exon 7 and intron 7 (NonSyn). c.577+1G>T, was localized in the donor splice site of introns 5 and 6 and led to transcripts lacking exon 5 (NonSyn).	[115] [116]
IFITM3	ISG which abrogates the release of IAV content from late endosomes into the cytoplasm. IFITM3 increases the survival of mouse lung-resident CD8$^+$ T cells after IAV infection, which can help clear the infection.	rs12252, leading to an alteration of the first splice acceptor site, leading to an IFITM3 protein lacking the first 21 amino acids (NonSyn). rs34481144, is located in the 5′-UTR and affects IFITM3 expression with the risk allele showing lower mRNA expression (NCR).	[100,118,120,122] [123]
IL-1B	Inflammatory cytokine involved in the development of adaptive immune responses. Furthermore, accumulating data has suggested that IL-1A and IL-1B have critical roles in innate immunity against viral infections.	rs1143627, located 31 base pairs upstream from the transcription start site, on the IL-1B promoter. This nucleotide change is located in a TATA-box motif of IL-1B, affecting the transcription activity of IL-1B (NCR).	[128,129]

Table 1. *Cont.*

Gene	Function	SNPs (Type)	References
IL-1A	Inflammatory cytokine that plays important roles in the development of adaptive immune responses. Moreover, multiple pieces of evidence have suggested that IL-1A and IL-1B play relevant roles in innate immunity against viral infections.	rs17561; A114S (NonSyn).	[126]
TNF-α	Pro-inflammatory cytokine which orchestrates the host´s defense.	rs361525, a minor allele (A) at position 238 (NCR).	[92,130,131]
CCR5	Cytokine receptor which has a role in mediating leukocyte migration in response to its ligands, including MIP-1a, MIP-1b, and RANTES. Furthermore, it can help direct many immune cell subsets, including regulatory T cells and Th17 cells to sites of infection, supporting the antiviral immune response.	CCR5-Δ32 allele (NonSyn).	[135]
CD55	Blocks C3 and C5 activation by inhibiting the formation of new C3 and C5 convertases, which are two proteases involved in complement activation and inflammation. CD55 functions to protect cells from complement attack and decreases the amplification of the complement cascade	rs2564978, resides in the minimal promoter region, affecting gene expression (NCR).	[47,79,100,137,138]
C1QBP	Binds to the globular heads of C1q molecules activating the classical pathway of complement.	rs3786054, localized in an intron (NCR).	[138,140]
SFTPA2	Soluble pattern-recognition molecule that may neutralize IAV infection.	-rs1965708; Q223K (NonSyn). -rs1059046; T9N (NonSyn).	[145]
Galectin-1	Recognizes galactose-containing oligosaccharides present in the cellular plasma membranes and in viruses, such as IAV.	-rs4820294 (NCR). -rs13057866 (NCR).	[147]
TMPRSS2	Type II transmembrane serine protease family member which activates HA proteins of diverse human IAV in tissue culture cells. Deletion of *Tmprss2* in mice impairs the spread of H1N1 influenza viruses, including the pH1N1. Moreover, body weight loss and survival were less severe in *Tmprss2* mutant mice compared to wild type mice after infection with H3N2 IAV.	-rs2070788, localized in an intron (NCR). -rs383510, localized in an intron (NCR).	[149]

Syn-synonymous, NonSyn-nonsynonymous, NCR-non-coding region (Intron, regulatory regions, promoter or UTR).

3. SNPs in Genes that Influence the IAV Vaccine Response

Currently, IAV vaccines are the main strategy to prevent IAV infection, though their effectiveness is suboptimal in many cases. Notably, the efficacy of vaccines against IAV infections can fluctuate and there is a significant immune response variability across the population. Factors such as previous exposure to IAV infections or vaccines, age, and the closeness of the match between the vaccine and circulating strains are important to explain differences in vaccine effectiveness between seasons and group populations [44,46,150–152]. However, multiple reports have demonstrated that the host genetic background and polymorphisms on key immune response genes modulate the immune response to infection or vaccination [153–160]. Therefore, new insights into IAV-host interaction and immune response modulating factors could allow us to design better vaccination strategies.

SNPs may modify the humoral immune response after IAV vaccination. Therefore, their impact on the immune responses induced after IAV vaccination are being analyzed [153–156]. The major histocompatibility complex (MHC) is localized in chromosome 6 of the human genome, it includes multiple genes and exhibits considerable diversity between populations. Moreover, in this genomic region, there is a higher presence of SNPs than in other sections of the genome. MHC class I and class II molecules have an essential role in the adaptive immune system in response to infections. Both classes of proteins bind peptide fragments derived from pathogens to be presented on the cell surface for recognition by appropriate T cells [97,161,162]. In those genes, the human leukocyte antigens (HLA) class I and II are important because of their role in the immune system. Gelder et al. studied whether HLA class II polymorphisms modulate anti-IAV antibody responses to vaccination in a United Kingdom population [154]. For that, a cohort of HLA-typed donors at risk was investigated, and hemagglutination-inhibition (HAI) titers were evaluated before and 28 days after the administration of seasonal trivalent influenza vaccine. A correlation between HLA class II alleles and IAV HAI titers in the influenza risk group was found. Moreover, a positive association between non-responsiveness to influenza vaccine and HLA-DRB1*07 and a negative association with HLA-DRB1*13 and HLA-DQB1*0603-9/14 [154] was reported, suggesting that polymorphisms in HLA class II molecules affect antibody responses to IAV vaccination. These findings are important because they could potentially identify individuals who may not be protected by current vaccination approaches.

In another study, Poland et al. analyzed the immunogenetic relationships between HLA, cytokine and cytokine receptor gene polymorphisms in the induction of antibodies in response to inactivated seasonal vaccines [156]. Authors did not find statistically significant associations between HLA class II alleles and IAV HAI titers. However, they established a positive association of some HLA class I alleles and IAV H1N1 HAI titers, including HLA- A*1101, A*6801, B*3503, B*1401, and C*0802. In contrast, they did not find associations between the HLA-A, B or C alleles and HAI antibody titers for IAV H3N2. In addition, when authors evaluated a panel of 586 cytokine and cytokine receptor SNPs, they identified several significant associations between SNPs, in regulatory or coding regions of cytokine (IL-6, IL-12B) or cytokine receptor (IL-1R, IL-10RB, TNFRSF1A) genes and variations in HAI antibody titers for IAV H1N1 [156] (Table 2). Notably, SNPs from three genes, IL-6 (rs1800796), IL-12B (rs3212227) and IL-1R1 (rs3732131) revealed links with IAV H1N1-induced antibody responses in an allele dose-related way. The presence of SNP allele C or G in the IL-12B or IL-1R1genes, respectively resulted in reduced HAI titers. However, high HAI titers in the presence of minor SNP allele G in the IL-6 gene were observed [156]. SNPs associations between cytokine or cytokine receptor genes and IAV H3N2 HAI titers were also identified (Table 2). For example, a variant GA for non-synonymous SNPs within the IL-12 receptor gene (rs2307153; D465G) and TNF receptor 2 gene (rs5746026; K232E) displayed associations with lower HAI titers, while a minor allele T variant (rs12722605) located in the 3′ region of the IL-2 receptor gene was related with high antibody titers (Table 2). These data suggest that host SNPs affect responses to influenza vaccine.

Mannose-binding lectin 2 (MBL-2) is a protein that binds N-acetylglucosamine, mannose, and fucose on different microorganisms and activates the lectin complement pathway [163,164]. Tang et al. studied the presence of SNPs in subjects who received an inactivated influenza vaccine. For that, authors classified the vaccine recipients in poor, normal or adverse responders. They observed that the G to A SNP in the codon 54 allele (rs1800450) in MBL-2 was associated with a decreased risk for the development of adverse or poor responses (Table 2) [165]. In addition, they did not find a significant association between responses and either TNF-α or IL-10 promoter SNPs among the 3 response groups [165].

Table 2. Associations between SNPs and IAV vaccine responses.

Gene	Function	SNPs (Type)	Vaccine	Reference
IL-6	Cytokine expressed as a response to infections or tissue injuries. It plays an important role in host defense through the stimulation of acute-phase responses.	-rs1800796 (NCR). - rs2069861 (NCR).	IIV	[156]
IL-12B	Cytokine that serves as a crucial inducer of Th1 cell development.	rs3212227, located in 3′UTR (NCR).	IIV	[156]
IFN-B1	Cytokine released as part of the innate immune response against infection by viruses or other pathogens.	rs1364613 (NCR).	IIV	[156]
TNFRSF1A	Cytokine receptor, its interaction with TNF-α control cell survival, apoptosis, and inflammation.	rs4149621 (NCR).	IIV	[156]
IL-1R1	Cytokine receptor involved in inflammatory and immune responses.	rs3732131, located in 3′UTR (NCR).	IIV	[156]
IL-10RB	Cytokine receptor that mediates the activation of the JAK/STAT signaling pathway leading to the expression of ISG.	rs3171425, located in 3′UTR (NCR).	IIV	[156]
IL-2RA	This cytokine receptor is important for the signaling pathway leading to immune cell differentiation and function.	-rs2228150 (Syn). -rs12722605 (NCR).	IIV	[156]
IL-10RA	Cytokine receptor that is involved in the inhibition of the synthesis of several proinflammatory cytokines.	-rs4252249 (Syn) -rs4252243 (NCR).	IIV	[156]
IL-12RB2	Cytokine receptor that plays a role in Th1 cell differentiation.	rs2307153; D465G (NonSyn).	IIV	[156]
IL-1RN	Cytokine receptor which modulates a variety of immune and inflammatory responses related with IL-1.	-rs315952 (Syn). -rs315951 located in 3′UTR (NCR).	IIV	[156]
TNFRSF1B	Cytokine receptor involved in the recruitment of anti-apoptotic proteins.	rs5746026; K232E (NonSyn)	IIV	[156]
MBL-2	This calcium-dependent protein that plays an important role in innate immunity, and activates the lectin complement pathway.	rs1800450; G54D (NonSyn)	IIV	[165]
IL-28B (IFNL3)	Type III IFN molecule, with brad functions in antiviral responses	rs8099917 (NCR).	IIV	[153]

Syn-synonymous, NonSyn-nonsynonymous, NCR-non-coding region (Intron, regulatory regions, promoter or UTR). IIV: inactivated seasonal vaccine. IL-6, interleukin 6. IL-12B, interleukin 12. IFN-B1, interferon beta 1 (IFNβ). TNFRSF1A, TNF receptor superfamily member 1A. IL-1R1, interleukin 1 receptor type 1. IL-10RB, interleukin 10 receptor subunit beta. IL-2RA, interleukin 2 receptor subunit alpha. IL-10RA, interleukin 10 receptor subunit alpha. IL-12RB2, interleukin 12 receptor subunit beta 2. IL-1RN, interleukin 1 receptor antagonist. TNFRSF1B, TNF receptor superfamily member 1B. MBL-2, mannose binding lectin 2. IL-28 B or IFNL3, interferon lambda 3.

Other SNPs that are not related with immune responses have been also linked to vaccine effectiveness. Egli et al. revealed that the presence of the T/G or G/G genotype (rs8099917, minor-allele) in IL-28B (IFNλ3), a type III IFN, was linked with increased seroconversion in recipients of an inactivated influenza vaccine (Table 2) [153]. Moreover, IAV-stimulated B- and T-cells from the minor-allele carriers exhibited increased HLA-DR and IL-4 expression, respectively. In addition, the expression of IL-28B, but not IL-28A or IL-29, mRNAs was significantly reduced in the rs8099917, minor-allele carriers.

Authors also reported that the IL-28B rs8099917 polymorphism affected humoral responses to the IAV vaccine, and had a strong outcome on cellular immune responses by modulating the Th1/Th2 cytokine response [153]. These findings are important because they will help to predict which individuals could not be protected by present vaccines and they can also be used to design personalized vaccine strategies to optimize the immune reaction.

4. Conclusions

The sequencing of the human genome together with the development of novel bioinformatic tools have made possible the identification of multiple SNPs. More information is available for the scientific community in the databases. In addition, the identification and study of the human genome variability has opened the opportunity to investigate their association with the risk of developing multiple human diseases facilitating their diagnosis or the susceptibility to infections caused by viruses or other pathogens. Moreover, the knowledge and analysis of genomic variability will be a valuable tool to predict the outcome of prophylactic or therapeutic interventions, including vaccines and drugs. The analysis of human SNPs and their association with IAV infections or vaccination outcomes have just begun. However, current research and data reflect the importance to obtain a better understanding of these relations and the mechanisms underlying the effect of SNPs in the human immune system. In the future, this knowledge could be used to better understand host factors affecting viral replication and disease severity and to develop new and more effective therapeutic strategies against viral infections.

Author Contributions: A.N. and M.L.D. conceptualized and wrote the manuscript.

Funding: This work was supported with funds from Comunidad de Madrid, Spain, reference 2017-T1/BMD-5155 and the Spanish Ministry of Science, Innovation and Universities (RTI-2018-094213-A-I00) to M.L.D and the "Ramon y Cajal" Fellowship RYC-2017 from Spanish Ministry of Economy, Industry and Competitiveness to A.N.

Acknowledgments: We apologize for those publications we could not refer due to space limitations.

Conflicts of Interest: The authors declare no conflict of interest.

References

1. Heldt, F.S.; Frensing, T.; Reichl, U. Modeling the intracellular dynamics of influenza virus replication to understand the control of viral RNA synthesis. *J. Virol.* **2012**, *86*, 7806–7817. [CrossRef] [PubMed]
2. Martinez-Sobrido, L.; Peersen, O.; Nogales, A. Temperature Sensitive Mutations in Influenza A Viral Ribonucleoprotein Complex Responsible for the Attenuation of the Live Attenuated Influenza Vaccine. *Viruses* **2018**, *10*, 560. [CrossRef] [PubMed]
3. Arranz, R.; Coloma, R.; Chichon, F.J.; Conesa, J.J.; Carrascosa, J.L.; Valpuesta, J.M.; Ortin, J.; Martin-Benito, J. The structure of native influenza virion ribonucleoproteins. *Science* **2012**, *338*, 1634–1637. [CrossRef]
4. Pflug, A.; Guilligay, D.; Reich, S.; Cusack, S. Structure of influenza A polymerase bound to the viral RNA promoter. *Nature* **2014**, *516*, 355–360. [CrossRef] [PubMed]
5. Pflug, A.; Lukarska, M.; Resa-Infante, P.; Reich, S.; Cusack, S. Structural insights into RNA synthesis by the influenza virus transcription-replication machine. *Virus Res.* **2017**, *234*, 103–117. [CrossRef] [PubMed]
6. Resa-Infante, P.; Jorba, N.; Coloma, R.; Ortin, J. The influenza virus RNA synthesis machine: Advances in its structure and function. *RNA Biol.* **2011**, *8*, 207–215. [CrossRef]
7. Baker, S.F.; Nogales, A.; Martinez-Sobrido, L. Downregulating viral gene expression: Codon usage bias manipulation for the generation of novel influenza A virus vaccines. *Future Virol.* **2015**, *10*, 715–730. [CrossRef]
8. Wise, H.M.; Foeglein, A.; Sun, J.; Dalton, R.M.; Patel, S.; Howard, W.; Anderson, E.C.; Barclay, W.S.; Digard, P. A complicated message: Identification of a novel PB1-related protein translated from influenza A virus segment 2 mRNA. *J. Virol.* **2009**, *83*, 8021–8031. [CrossRef]
9. Jagger, B.W.; Wise, H.M.; Kash, J.C.; Walters, K.A.; Wills, N.M.; Xiao, Y.L.; Dunfee, R.L.; Schwartzman, L.M.; Ozinsky, A.; Bell, G.L.; et al. An overlapping protein-coding region in influenza A virus segment 3 modulates the host response. *Science* **2012**, *337*, 199–204. [CrossRef]
10. Bouvier, N.M.; Palese, P. The biology of influenza viruses. *Vaccine* **2008**, *26*, D49–D53. [CrossRef]

11. Byrd-Leotis, L.; Cummings, R.D.; Steinhauer, D.A. The Interplay between the Host Receptor and Influenza Virus Hemagglutinin and Neuraminidase. *Int. J. Mol. Sci.* **2017**, *18*, 1541. [CrossRef] [PubMed]
12. De, A. Molecular evolution of hemagglutinin gene of Influenza A virus. *Front. Biosci.* **2018**, *10*, 101–118. [CrossRef]
13. Imai, M.; Kawaoka, Y. The role of receptor binding specificity in interspecies transmission of influenza viruses. *Curr. Opin. Virol.* **2012**, *2*, 160–167. [CrossRef] [PubMed]
14. McAuley, J.L.; Gilbertson, B.P.; Trifkovic, S.; Brown, L.E.; McKimm-Breschkin, J.L. Influenza Virus Neuraminidase Structure and Functions. *Front. Microbiol.* **2019**, *10*, 39. [CrossRef] [PubMed]
15. Gaymard, A.; Le Briand, N.; Frobert, E.; Lina, B.; Escuret, V. Functional balance between neuraminidase and haemagglutinin in influenza viruses. *Clin. Microbiol. Infect.* **2016**, *22*, 975–983. [CrossRef] [PubMed]
16. Jagadesh, A.; Salam, A.A.; Mudgal, P.P.; Arunkumar, G. Influenza virus neuraminidase (NA): A target for antivirals and vaccines. *Arch. Virol.* **2016**, *161*, 2087–2094. [CrossRef]
17. Chen, B.J.; Leser, G.P.; Jackson, D.; Lamb, R.A. The influenza virus M2 protein cytoplasmic tail interacts with the M1 protein and influences virus assembly at the site of virus budding. *J. Virol.* **2008**, *82*, 10059–10070. [CrossRef] [PubMed]
18. Rossman, J.S.; Lamb, R.A. Influenza virus assembly and budding. *Virology* **2011**, *411*, 229–236. [CrossRef]
19. Chen, B.J.; Leser, G.P.; Morita, E.; Lamb, R.A. Influenza virus hemagglutinin and neuraminidase, but not the matrix protein, are required for assembly and budding of plasmid-derived virus-like particles. *J. Virol.* **2007**, *81*, 7111–7123. [CrossRef]
20. Nayak, D.P.; Hui, E.K.; Barman, S. Assembly and budding of influenza virus. *Virus Res.* **2004**, *106*, 147–165. [CrossRef]
21. Ward, A.C.; Castelli, L.A.; Lucantoni, A.C.; White, J.F.; Azad, A.A.; Macreadie, I.G. Expression and analysis of the NS2 protein of influenza A virus. *Arch. Virol.* **1995**, *140*, 2067–2073. [CrossRef] [PubMed]
22. Yasuda, J.; Nakada, S.; Kato, A.; Toyoda, T.; Ishihama, A. Molecular assembly of influenza virus: Association of the NS2 protein with virion matrix. *Virology* **1993**, *196*, 249–255. [CrossRef] [PubMed]
23. Richardson, J.C.; Akkina, R.K. NS2 protein of influenza virus is found in purified virus and phosphorylated in infected cells. *Arch. Virol.* **1991**, *116*, 69–80. [CrossRef] [PubMed]
24. Paterson, D.; Fodor, E. Emerging roles for the influenza A virus nuclear export protein (NEP). *PLoS Pathog.* **2012**, *8*, e1003019. [CrossRef] [PubMed]
25. Nogales, A.; Martinez-Sobrido, L.; Topham, D.J.; DeDiego, M.L. Modulation of Innate Immune Responses by the Influenza A NS1 and PA-X Proteins. *Viruses* **2018**, *10*, 708. [CrossRef]
26. Nogales, A.; Rodriguez, L.; DeDiego, M.L.; Topham, D.J.; Martinez-Sobrido, L. Interplay of PA-X and NS1 Proteins in Replication and Pathogenesis of a Temperature-Sensitive 2009 Pandemic H1N1 Influenza A Virus. *J. Virol.* **2017**, *91*, e00720-17. [CrossRef] [PubMed]
27. Gao, H.; Sun, H.; Hu, J.; Qi, L.; Wang, J.; Xiong, X.; Wang, Y.; He, Q.; Lin, Y.; Kong, W.; et al. Twenty amino acids at the C-terminus of PA-X are associated with increased influenza A virus replication and pathogenicity. *J. Gen. Virol.* **2015**, *96*, 2036–2049. [CrossRef]
28. Hu, J.; Mo, Y.; Gao, Z.; Wang, X.; Gu, M.; Liang, Y.; Cheng, X.; Hu, S.; Liu, W.; Liu, H.; et al. PA-X-associated early alleviation of the acute lung injury contributes to the attenuation of a highly pathogenic H5N1 avian influenza virus in mice. *Med. Microbiol. Immunol.* **2016**, *205*, 381–395. [CrossRef] [PubMed]
29. Hu, J.; Mo, Y.; Wang, X.; Gu, M.; Hu, Z.; Zhong, L.; Wu, Q.; Hao, X.; Hu, S.; Liu, W.; et al. PA-X decreases the pathogenicity of highly pathogenic H5N1 influenza A virus in avian species by inhibiting virus replication and host response. *J. Virol.* **2015**, *89*, 4126–4142. [CrossRef]
30. Khaperskyy, D.A.; Schmaling, S.; Larkins-Ford, J.; McCormick, C.; Gaglia, M.M. Selective Degradation of Host RNA Polymerase II Transcripts by Influenza A Virus PA-X Host Shutoff Protein. *PLoS Pathog.* **2016**, *12*, e1005427. [CrossRef]
31. Barr, I.G.; McCauley, J.; Cox, N.; Daniels, R.; Engelhardt, O.G.; Fukuda, K.; Grohmann, G.; Hay, A.; Kelso, A.; Klimov, A.; et al. Epidemiological, antigenic and genetic characteristics of seasonal influenza A(H1N1), A(H3N2) and B influenza viruses: Basis for the WHO recommendation on the composition of influenza vaccines for use in the 2009-2010 northern hemisphere season. *Vaccine* **2010**, *28*, 1156–1167. [CrossRef] [PubMed]
32. Grohskopf, L.A.; Sokolow, L.Z.; Broder, K.R.; Olsen, S.J.; Karron, R.A.; Jernigan, D.B.; Bresee, J.S. Prevention and Control of Seasonal Influenza with Vaccines. *MMWR Recomm. Rep.* **2016**, *65*, 1–54. [CrossRef] [PubMed]

33. Molinari, N.A.; Ortega-Sanchez, I.R.; Messonnier, M.L.; Thompson, W.W.; Wortley, P.M.; Weintraub, E.; Bridges, C.B. The annual impact of seasonal influenza in the US: Measuring disease burden and costs. *Vaccine* **2007**, *25*, 5086–5096. [CrossRef] [PubMed]
34. Bailey, E.S.; Choi, J.Y.; Fieldhouse, J.K.; Borkenhagen, L.K.; Zemke, J.; Zhang, D.; Gray, G.C. The continual threat of influenza virus infections at the human-animal interface: What is new from a one health perspective? *Evol. Med. Public Health* **2018**, *2018*, 192–198. [CrossRef] [PubMed]
35. Parrish, C.R.; Murcia, P.R.; Holmes, E.C. Influenza virus reservoirs and intermediate hosts: Dogs, horses, and new possibilities for influenza virus exposure of humans. *J. Virol.* **2015**, *89*, 2990–2994. [CrossRef] [PubMed]
36. Tong, S.; Zhu, X.; Li, Y.; Shi, M.; Zhang, J.; Bourgeois, M.; Yang, H.; Chen, X.; Recuenco, S.; Gomez, J.; et al. New world bats harbor diverse influenza A viruses. *PLoS Pathog.* **2013**, *9*, e1003657. [CrossRef] [PubMed]
37. Yoon, S.W.; Webby, R.J.; Webster, R.G. Evolution and ecology of influenza A viruses. *Curr. Top. Microbiol. Immunol.* **2014**, *385*, 359–375. [CrossRef] [PubMed]
38. Clark, A.M.; DeDiego, M.L.; Anderson, C.S.; Wang, J.; Yang, H.; Nogales, A.; Martinez-Sobrido, L.; Zand, M.S.; Sangster, M.Y.; Topham, D.J. Antigenicity of the 2015-2016 seasonal H1N1 human influenza virus HA and NA proteins. *PLoS ONE* **2017**, *12*, e0188267. [CrossRef]
39. Grohskopf, L.A.; Sokolow, L.Z.; Broder, K.R.; Walter, E.B.; Bresee, J.S.; Fry, A.M.; Jernigan, D.B. Prevention and Control of Seasonal Influenza with Vaccines: Recommendations of the Advisory Committee on Immunization Practices—United States, 2017–2018 Influenza Season. *MMWR Recomm. Rep.* **2017**, *66*, 1–20. [CrossRef] [PubMed]
40. Garten, R.J.; Davis, C.T.; Russell, C.A.; Shu, B.; Lindstrom, S.; Balish, A.; Sessions, W.M.; Xu, X.; Skepner, E.; Deyde, V.; et al. Antigenic and genetic characteristics of swine-origin 2009 A(H1N1) influenza viruses circulating in humans. *Science* **2009**, *325*, 197–201. [CrossRef]
41. Smith, G.J.; Vijaykrishna, D.; Bahl, J.; Lycett, S.J.; Worobey, M.; Pybus, O.G.; Ma, S.K.; Cheung, C.L.; Raghwani, J.; Bhatt, S.; et al. Origins and evolutionary genomics of the 2009 swine-origin H1N1 influenza A epidemic. *Nature* **2009**, *459*, 1122–1125. [CrossRef] [PubMed]
42. Baker, S.F.; Nogales, A.; Santiago, F.W.; Topham, D.J.; Martinez-Sobrido, L. Competitive detection of influenza neutralizing antibodies using a novel bivalent fluorescence-based microneutralization assay (BiFMA). *Vaccine* **2015**, *33*, 3562–3570. [CrossRef] [PubMed]
43. Nogales, A.; Martinez-Sobrido, L. Reverse Genetics Approaches for the Development of Influenza Vaccines. *Int. J. Mol. Sci.* **2016**, *18*, 20. [CrossRef] [PubMed]
44. Schotsaert, M.; Garcia-Sastre, A. Inactivated influenza virus vaccines: The future of TIV and QIV. *Curr. Opin. Virol.* **2017**, *23*, 102–106. [CrossRef] [PubMed]
45. Jackson, R.J.; Cooper, K.L.; Tappenden, P.; Rees, A.; Simpson, E.L.; Read, R.C.; Nicholson, K.G. Oseltamivir, zanamivir and amantadine in the prevention of influenza: A systematic review. *J. Infect.* **2011**, *62*, 14–25. [CrossRef] [PubMed]
46. Blanco-Lobo, P.; Nogales, A.; Rodriguez, L.; Martinez-Sobrido, L. Novel Approaches for The Development of Live Attenuated Influenza Vaccines. *Viruses* **2019**, *11*, 190. [CrossRef] [PubMed]
47. Lin, T.Y.; Brass, A.L. Host genetic determinants of influenza pathogenicity. *Curr. Opin. Virol.* **2013**, *3*, 531–536. [CrossRef] [PubMed]
48. Bui, C.M.; Chughtai, A.A.; Adam, D.C.; MacIntyre, C.R. An overview of the epidemiology and emergence of influenza A infection in humans over time. *Arch. Public Health* **2017**, *75*, 15. [CrossRef] [PubMed]
49. Perez, D.R.; Garcia-Sastre, A. H5N1, a wealth of knowledge to improve pandemic preparedness. *Virus Res.* **2013**, *178*, 1–2. [CrossRef] [PubMed]
50. Iwasaki, A.; Pillai, P.S. Innate immunity to influenza virus infection. *Nat. Rev. Immunol.* **2014**, *14*, 315–328. [CrossRef] [PubMed]
51. Goff, P.H.; Hayashi, T.; He, W.; Yao, S.; Cottam, H.B.; Tan, G.S.; Crain, B.; Krammer, F.; Messer, K.; Pu, M.; et al. Synthetic Toll-Like Receptor 4 (TLR4) and TLR7 Ligands Work Additively via MyD88 To Induce Protective Antiviral Immunity in Mice. *J. Virol.* **2017**, *91*. [CrossRef] [PubMed]
52. Weber-Gerlach, M.; Weber, F. Standing on three legs: Antiviral activities of RIG-I against influenza viruses. *Curr. Opin. Immunol.* **2016**, *42*, 71–75. [CrossRef] [PubMed]
53. Sarvestani, S.T.; McAuley, J.L. The role of the NLRP3 inflammasome in regulation of antiviral responses to influenza A virus infection. *Antivir. Res.* **2017**, *148*, 32–42. [CrossRef] [PubMed]

54. Zhang, H.; Luo, J.; Alcorn, J.F.; Chen, K.; Fan, S.; Pilewski, J.; Liu, A.; Chen, W.; Kolls, J.K.; Wang, J. AIM2 Inflammasome Is Critical for Influenza-Induced Lung Injury and Mortality. *J. Immunol.* **2017**, *198*, 4383–4393. [CrossRef] [PubMed]
55. Wolff, T.; Ludwig, S. Influenza viruses control the vertebrate type I interferon system: Factors, mechanisms, and consequences. *J. Interferon Cytokine Res.* **2009**, *29*, 549–557. [CrossRef] [PubMed]
56. Hermant, P.; Michiels, T. Interferon-lambda in the context of viral infections: Production, response and therapeutic implications. *J. Innate Immun.* **2014**, *6*, 563–574. [CrossRef] [PubMed]
57. Mogensen, T.H.; Paludan, S.R. Molecular pathways in virus-induced cytokine production. *Microbiol. Mol. Biol. Rev.* **2001**, *65*, 131–150. [CrossRef]
58. Crotta, S.; Davidson, S.; Mahlakoiv, T.; Desmet, C.J.; Buckwalter, M.R.; Albert, M.L.; Staeheli, P.; Wack, A. Type I and type III interferons drive redundant amplification loops to induce a transcriptional signature in influenza-infected airway epithelia. *PLoS Pathog.* **2013**, *9*, e1003773. [CrossRef] [PubMed]
59. Desai, T.M.; Marin, M.; Chin, C.R.; Savidis, G.; Brass, A.L.; Melikyan, G.B. IFITM3 restricts influenza A virus entry by blocking the formation of fusion pores following virus-endosome hemifusion. *PLoS Pathog.* **2014**, *10*, e1004048. [CrossRef]
60. Levy, D.E.; Marie, I.J.; Durbin, J.E. Induction and function of type I and III interferon in response to viral infection. *Curr. Opin. Virol.* **2011**, *1*, 476–486. [CrossRef]
61. Schindler, C.; Levy, D.E.; Decker, T. JAK-STAT signaling: From interferons to cytokines. *J. Biol. Chem.* **2007**, *282*, 20059–20063. [CrossRef] [PubMed]
62. Tenoever, B.R.; Ng, S.L.; Chua, M.A.; McWhirter, S.M.; Garcia-Sastre, A.; Maniatis, T. Multiple functions of the IKK-related kinase IKKepsilon in interferon-mediated antiviral immunity. *Science* **2007**, *315*, 1274–1278. [CrossRef] [PubMed]
63. Ng, S.L.; Friedman, B.A.; Schmid, S.; Gertz, J.; Myers, R.M.; tenOever, B.R.; Maniatis, T. I kappa B kinase epsilon (IKK epsilon) regulates the balance between type I and type II interferon responses. *Proc. Natl. Acad. Sci. USA* **2011**, *108*, 21170–21175. [CrossRef] [PubMed]
64. Hoffmann, H.H.; Schneider, W.M.; Rice, C.M. Interferons and viruses: An evolutionary arms race of molecular interactions. *Trends Immunol.* **2015**, *36*, 124–138. [CrossRef] [PubMed]
65. Kos, F.J.; Engleman, E.G. Role of natural killer cells in the generation of influenza virus-specific cytotoxic T cells. *Cell. Immunol.* **1996**, *173*, 1–6. [CrossRef] [PubMed]
66. He, X.S.; Draghi, M.; Mahmood, K.; Holmes, T.H.; Kemble, G.W.; Dekker, C.L.; Arvin, A.M.; Parham, P.; Greenberg, H.B. T cell-dependent production of IFN-gamma by NK cells in response to influenza A virus. *J. Clin. Investig.* **2004**, *114*, 1812–1819. [CrossRef]
67. Hashimoto, Y.; Moki, T.; Takizawa, T.; Shiratsuchi, A.; Nakanishi, Y. Evidence for phagocytosis of influenza virus-infected, apoptotic cells by neutrophils and macrophages in mice. *J. Immunol.* **2007**, *178*, 2448–2457. [CrossRef]
68. Legge, K.L.; Braciale, T.J. Accelerated migration of respiratory dendritic cells to the regional lymph nodes is limited to the early phase of pulmonary infection. *Immunity* **2003**, *18*, 265–277. [CrossRef]
69. Legge, K.L.; Braciale, T.J. Lymph node dendritic cells control CD8+ T cell responses through regulated FasL expression. *Immunity* **2005**, *23*, 649–659. [CrossRef]
70. Chiu, C.; Openshaw, P.J. Antiviral B cell and T cell immunity in the lungs. *Nat. Immunol.* **2015**, *16*, 18–26. [CrossRef]
71. Guarda, G.; Zenger, M.; Yazdi, A.S.; Schroder, K.; Ferrero, I.; Menu, P.; Tardivel, A.; Mattmann, C.; Tschopp, J. Differential expression of NLRP3 among hematopoietic cells. *J. Immunol.* **2011**, *186*, 2529–2534. [CrossRef] [PubMed]
72. Pothlichet, J.; Meunier, I.; Davis, B.K.; Ting, J.P.; Skamene, E.; von Messling, V.; Vidal, S.M. Type I IFN triggers RIG-I/TLR3/NLRP3-dependent inflammasome activation in influenza A virus infected cells. *PLoS Pathog.* **2013**, *9*, e1003256. [CrossRef] [PubMed]
73. Lupfer, C.; Kanneganti, T.D. The expanding role of NLRs in antiviral immunity. *Immunol. Rev.* **2013**, *255*, 13–24. [CrossRef] [PubMed]
74. Allen, I.C.; Scull, M.A.; Moore, C.B.; Holl, E.K.; McElvania-TeKippe, E.; Taxman, D.J.; Guthrie, E.H.; Pickles, R.J.; Ting, J.P. The NLRP3 inflammasome mediates in vivo innate immunity to influenza A virus through recognition of viral RNA. *Immunity* **2009**, *30*, 556–565. [CrossRef] [PubMed]

75. McAuley, J.L.; Tate, M.D.; MacKenzie-Kludas, C.J.; Pinar, A.; Zeng, W.; Stutz, A.; Latz, E.; Brown, L.E.; Mansell, A. Activation of the NLRP3 inflammasome by IAV virulence protein PB1-F2 contributes to severe pathophysiology and disease. *PLoS Pathog.* **2013**, *9*, e1003392. [CrossRef] [PubMed]
76. Ichinohe, T.; Pang, I.K.; Iwasaki, A. Influenza virus activates inflammasomes via its intracellular M2 ion channel. *Nat. Immunol.* **2010**, *11*, 404–410. [CrossRef] [PubMed]
77. Agrawal, P.; Nawadkar, R.; Ojha, H.; Kumar, J.; Sahu, A. Complement Evasion Strategies of Viruses: An Overview. *Front. Microbiol.* **2017**, *8*, 1117. [CrossRef]
78. Kishore, U.; Reid, K.B. C1q: Structure, function, and receptors. *Immunopharmacology* **2000**, *49*, 159–170. [CrossRef]
79. Kim, D.D.; Song, W.C. Membrane complement regulatory proteins. *Clin. Immunol.* **2006**, *118*, 127–136. [CrossRef]
80. Wang, R.; Xiao, H.; Guo, R.; Li, Y.; Shen, B. The role of C5a in acute lung injury induced by highly pathogenic viral infections. *Emerg Microbes Infect* **2015**, *4*, e28. [CrossRef]
81. Kwok, P.Y.; Chen, X. Detection of single nucleotide polymorphisms. *Curr. Issues Mol. Biol.* **2003**, *5*, 43–60. [PubMed]
82. Ramirez-Bello, J.; Jimenez-Morales, M. Functional implications of single nucleotide polymorphisms (SNPs) in protein-coding and non-coding RNA genes in multifactorial diseases. *Gac. Med. Mex.* **2017**, *153*, 238–250. [PubMed]
83. Ramirez-Bello, J.; Vargas-Alarcon, G.; Tovilla-Zarate, C.; Fragoso, J.M. Single nucleotide polymorphisms (SNPs): Functional implications of regulatory-SNP (rSNP) and structural RNA (srSNPs) in complex diseases. *Gac. Med. Mex.* **2013**, *149*, 220–228. [PubMed]
84. Skevaki, C.; Pararas, M.; Kostelidou, K.; Tsakris, A.; Routsias, J.G. Single nucleotide polymorphisms of Toll-like receptors and susceptibility to infectious diseases. *Clin. Exp. Immunol.* **2015**, *180*, 165–177. [CrossRef] [PubMed]
85. Hill, A.V. The genomics and genetics of human infectious disease susceptibility. *Annu. Rev. Genom. Hum. Genet.* **2001**, *2*, 373–400. [CrossRef]
86. Dai, W.; Ye, Z.; Lu, H.; Su, Q.; Li, H.; Li, L. Meta-analysis of the relationship between single nucleotide polymorphism of IL-10-1082G/A and rheumatic heart disease. *Oncotarget* **2018**, *9*, 12343–12350. [CrossRef] [PubMed]
87. Koberle, B.; Koch, B.; Fischer, B.M.; Hartwig, A. Single nucleotide polymorphisms in DNA repair genes and putative cancer risk. *Arch Toxicol* **2016**, *90*, 2369–2388. [CrossRef]
88. Li, X.; Zhou, J.; Chen, H.; Wang, F.; Mei, Q.; Sun, H. The association between the UBQLN1 polymorphism and Alzheimer's disease risk: A systematic review. *Cell. Mol. Biol.* **2017**, *63*, 94–96. [CrossRef]
89. Luykx, J.J.; Broersen, J.L.; de Leeuw, M. The DRD2 rs1076560 polymorphism and schizophrenia-related intermediate phenotypes: A systematic review and meta-analysis. *Neurosci. Biobehav. Rev.* **2017**, *74*, 214–224. [CrossRef]
90. Shaw, V.; Bullock, K.; Greenhalf, W. Single-Nucleotide Polymorphism to Associate Cancer Risk. *Methods Mol. Biol.* **2016**, *1381*, 93–110. [CrossRef]
91. Zhang, Y.; Bai, R.; Liu, C.; Ma, C.; Chen, X.; Yang, J.; Sun, D. MicroRNA single-nucleotide polymorphisms and diabetes mellitus: A comprehensive review. *Clin. Genet.* **2019**, *95*, 451–461. [CrossRef] [PubMed]
92. Imahara, S.D.; O'Keefe, G.E. Genetic determinants of the inflammatory response. *Curr. Opin. Crit. Care* **2004**, *10*, 318–324. [CrossRef] [PubMed]
93. Carvalho, T.L.; Lima, R.E.; Goes, G.H.B.; Pereira, L.A.; Fernandes, M.S.S.; Moura, P.; Vasconcelos, L.R.S.; Correia, C.C. Cognitive Dysfunction and Single Nucleotide Polymorphisms in Hepatitis C Virus-Infected Persons: A Systematic Review. *Viral Immunol.* **2017**, *30*, 703–707. [CrossRef] [PubMed]
94. Jaeger, M.; Stappers, M.H.; Joosten, L.A.; Gyssens, I.C.; Netea, M.G. Genetic variation in pattern recognition receptors: Functional consequences and susceptibility to infectious disease. *Future Med.* **2015**, *10*. [CrossRef] [PubMed]
95. Kenney, A.D.; Dowdle, J.A.; Bozzacco, L.; McMichael, T.M.; St Gelais, C.; Panfil, A.R.; Sun, Y.; Schlesinger, L.S.; Anderson, M.Z.; Green, P.L.; et al. Human Genetic Determinants of Viral Diseases. *Annu. Rev. Genet.* **2017**, *51*, 241–263. [CrossRef] [PubMed]
96. Mathew, S.; Abdel-Hafiz, H.; Raza, A.; Fatima, K.; Qadri, I. Host nucleotide polymorphism in hepatitis B virus-associated hepatocellular carcinoma. *World J. Hepatol.* **2016**, *8*, 485–498. [CrossRef] [PubMed]

97. Rehman, S.U.; Rauf, M.; Abbas, Z.; Hamed, M.H.; Qadri, I. Role of Some Predominant Host Immunomodulators' Single Nucleotide Polymorphisms in Severity of Hepatitis B Virus and Hepatitis C Virus Infection. *Viral Immunol.* **2016**, *29*, 536–545. [CrossRef] [PubMed]
98. Hidaka, F.; Matsuo, S.; Muta, T.; Takeshige, K.; Mizukami, T.; Nunoi, H. A missense mutation of the Toll-like receptor 3 gene in a patient with influenza-associated encephalopathy. *Clin. Immunol.* **2006**, *119*, 188–194. [CrossRef]
99. Esposito, S.; Molteni, C.G.; Giliani, S.; Mazza, C.; Scala, A.; Tagliaferri, L.; Pelucchi, C.; Fossali, E.; Plebani, A.; Principi, N. Toll-like receptor 3 gene polymorphisms and severity of pandemic A/H1N1/2009 influenza in otherwise healthy children. *Virol J.* **2012**, *9*, 270. [CrossRef]
100. Lee, N.; Cao, B.; Ke, C.; Lu, H.; Hu, Y.; Tam, C.H.T.; Ma, R.C.W.; Guan, D.; Zhu, Z.; Li, H.; et al. IFITM3, TLR3, and CD55 Gene SNPs and Cumulative Genetic Risks for Severe Outcomes in Chinese Patients With H7N9/H1N1pdm09 Influenza. *J. Infect. Dis.* **2017**, *216*, 97–104. [CrossRef]
101. Her, Z.; Teng, T.S.; Tan, J.J.; Teo, T.H.; Kam, Y.W.; Lum, F.M.; Lee, W.W.; Gabriel, C.; Melchiotti, R.; Andiappan, A.K.; et al. Loss of TLR3 aggravates CHIKV replication and pathology due to an altered virus-specific neutralizing antibody response. *EMBO Mol. Med.* **2015**, *7*, 24–41. [CrossRef] [PubMed]
102. Zhang, S.Y.; Jouanguy, E.; Ugolini, S.; Smahi, A.; Elain, G.; Romero, P.; Segal, D.; Sancho-Shimizu, V.; Lorenzo, L.; Puel, A.; et al. TLR3 deficiency in patients with herpes simplex encephalitis. *Science* **2007**, *317*, 1522–1527. [CrossRef] [PubMed]
103. Lafaille, F.G.; Pessach, I.M.; Zhang, S.Y.; Ciancanelli, M.J.; Herman, M.; Abhyankar, A.; Ying, S.W.; Keros, S.; Goldstein, P.A.; Mostoslavsky, G.; et al. Impaired intrinsic immunity to HSV-1 in human iPSC-derived TLR3-deficient CNS cells. *Nature* **2012**, *491*, 769–773. [CrossRef] [PubMed]
104. Perez de Diego, R.; Sancho-Shimizu, V.; Lorenzo, L.; Puel, A.; Plancoulaine, S.; Picard, C.; Herman, M.; Cardon, A.; Durandy, A.; Bustamante, J.; et al. Human TRAF3 adaptor molecule deficiency leads to impaired Toll-like receptor 3 response and susceptibility to herpes simplex encephalitis. *Immunity* **2010**, *33*, 400–411. [CrossRef] [PubMed]
105. Sancho-Shimizu, V.; Perez de Diego, R.; Lorenzo, L.; Halwani, R.; Alangari, A.; Israelsson, E.; Fabrega, S.; Cardon, A.; Maluenda, J.; Tatematsu, M.; et al. Herpes simplex encephalitis in children with autosomal recessive and dominant TRIF deficiency. *J. Clin. Investig.* **2011**, *121*, 4889–4902. [CrossRef] [PubMed]
106. Herman, M.; Ciancanelli, M.; Ou, Y.H.; Lorenzo, L.; Klaudel-Dreszler, M.; Pauwels, E.; Sancho-Shimizu, V.; Perez de Diego, R.; Abhyankar, A.; Israelsson, E.; et al. Heterozygous TBK1 mutations impair TLR3 immunity and underlie herpes simplex encephalitis of childhood. *J. Exp. Med.* **2012**, *209*, 1567–1582. [CrossRef]
107. Al-Anazi, M.R.; Matou-Nasri, S.; Abdo, A.A.; Sanai, F.M.; Alkahtani, S.; Alarifi, S.; Alkahtane, A.A.; Al-Yahya, H.; Ali, D.; Alessia, M.S.; et al. Association of Toll-Like Receptor 3 Single-Nucleotide Polymorphisms and Hepatitis C Virus Infection. *J. Immunol. Res.* **2017**, *2017*, 1590653. [CrossRef]
108. Al-Qahtani, A.; Al-Ahdal, M.; Abdo, A.; Sanai, F.; Al-Anazi, M.; Khalaf, N.; Viswan, N.A.; Al-Ashgar, H.; Al-Humaidan, H.; Al-Suwayeh, R.; et al. Toll-like receptor 3 polymorphism and its association with hepatitis B virus infection in Saudi Arabian patients. *J. Med. Virol.* **2012**, *84*, 1353–1359. [CrossRef]
109. Jorgensen, S.E.; Christiansen, M.; Ryo, L.B.; Gad, H.H.; Gjedsted, J.; Staeheli, P.; Mikkelsen, J.G.; Storgaard, M.; Hartmann, R.; Mogensen, T.H. Defective RNA sensing by RIG-I in severe influenza virus infection. *Clin. Exp. Immunol.* **2018**, *192*, 366–376. [CrossRef]
110. Honda, K.; Yanai, H.; Negishi, H.; Asagiri, M.; Sato, M.; Mizutani, T.; Shimada, N.; Ohba, Y.; Takaoka, A.; Yoshida, N.; et al. IRF-7 is the master regulator of type-I interferon-dependent immune responses. *Nature* **2005**, *434*, 772–777. [CrossRef]
111. Marie, I.; Durbin, J.E.; Levy, D.E. Differential viral induction of distinct interferon-alpha genes by positive feedback through interferon regulatory factor-7. *EMBO J.* **1998**, *17*, 6660–6669. [CrossRef] [PubMed]
112. Osterlund, P.I.; Pietila, T.E.; Veckman, V.; Kotenko, S.V.; Julkunen, I. IFN regulatory factor family members differentially regulate the expression of type III IFN (IFN-lambda) genes. *J. Immunol.* **2007**, *179*, 3434–3442. [CrossRef] [PubMed]
113. Ciancanelli, M.J.; Huang, S.X.; Luthra, P.; Garner, H.; Itan, Y.; Volpi, S.; Lafaille, F.G.; Trouillet, C.; Schmolke, M.; Albrecht, R.A.; et al. Infectious disease. Life-threatening influenza and impaired interferon amplification in human IRF7 deficiency. *Science* **2015**, *348*, 448–453. [CrossRef] [PubMed]

114. Chang, J.; Lindsay, R.J.; Kulkarni, S.; Lifson, J.D.; Carrington, M.; Altfeld, M. Polymorphisms in interferon regulatory factor 7 reduce interferon-alpha responses of plasmacytoid dendritic cells to HIV-1. *AIDS* **2011**, *25*, 715–717. [CrossRef] [PubMed]
115. Hernandez, N.; Melki, I.; Jing, H.; Habib, T.; Huang, S.S.Y.; Danielson, J.; Kula, T.; Drutman, S.; Belkaya, S.; Rattina, V.; et al. Life-threatening influenza pneumonitis in a child with inherited IRF9 deficiency. *J. Exp. Med.* **2018**, *215*, 2567–2585. [CrossRef]
116. Bravo Garcia-Morato, M.; Calvo Apalategi, A.; Bravo-Gallego, L.Y.; Blazquez Moreno, A.; Simon-Fuentes, M.; Garmendia, J.V.; Mendez Echevarria, A.; Del Rosal Rabes, T.; Dominguez-Soto, A.; Lopez-Granados, E.; et al. Impaired control of multiple viral infections in a family with complete IRF9 deficiency. *J. Allergy Clin. Immunol.* **2019**, *144*, 309–312. [CrossRef] [PubMed]
117. Wakim, L.M.; Gupta, N.; Mintern, J.D.; Villadangos, J.A. Enhanced survival of lung tissue-resident memory CD8(+) T cells during infection with influenza virus due to selective expression of IFITM3. *Nat. Immunol.* **2013**, *14*, 238–245. [CrossRef]
118. Everitt, A.R.; Clare, S.; Pertel, T.; John, S.P.; Wash, R.S.; Smith, S.E.; Chin, C.R.; Feeley, E.M.; Sims, J.S.; Adams, D.J.; et al. IFITM3 restricts the morbidity and mortality associated with influenza. *Nature* **2012**, *484*, 519–523. [CrossRef]
119. Weidner, J.M.; Jiang, D.; Pan, X.B.; Chang, J.; Block, T.M.; Guo, J.T. Interferon-induced cell membrane proteins, IFITM3 and tetherin, inhibit vesicular stomatitis virus infection via distinct mechanisms. *J. Virol.* **2010**, *84*, 12646–12657. [CrossRef]
120. Zhang, Y.H.; Zhao, Y.; Li, N.; Peng, Y.C.; Giannoulatou, E.; Jin, R.H.; Yan, H.P.; Wu, H.; Liu, J.H.; Liu, N.; et al. Interferon-induced transmembrane protein-3 genetic variant rs12252-C is associated with severe influenza in Chinese individuals. *Nat. Commun.* **2013**, *4*, 1418. [CrossRef]
121. Yang, X.; Tan, B.; Zhou, X.; Xue, J.; Zhang, X.; Wang, P.; Shao, C.; Li, Y.; Li, C.; Xia, H.; et al. Interferon-Inducible Transmembrane Protein 3 Genetic Variant rs12252 and Influenza Susceptibility and Severity: A Meta-Analysis. *PLoS ONE* **2015**, *10*, e0124985. [CrossRef] [PubMed]
122. Prabhu, S.S.; Chakraborty, T.T.; Kumar, N.; Banerjee, I. Association between IFITM3 rs12252 polymorphism and influenza susceptibility and severity: A meta-analysis. *Gene* **2018**, *674*, 70–79. [CrossRef] [PubMed]
123. Allen, E.K.; Randolph, A.G.; Bhangale, T.; Dogra, P.; Ohlson, M.; Oshansky, C.M.; Zamora, A.E.; Shannon, J.P.; Finkelstein, D.; Dressen, A.; et al. SNP-mediated disruption of CTCF binding at the IFITM3 promoter is associated with risk of severe influenza in humans. *Nat. Med.* **2017**, *23*, 975–983. [CrossRef]
124. Eisfeld, A.J.; Kawaoka, Y. Calculated risk: A new single-nucleotide polymorphism linked to severe influenza disease. *Nat. Med.* **2017**, *23*, 911–912. [CrossRef] [PubMed]
125. Dinarello, C.A. IL-1: Discoveries, controversies and future directions. *Eur. J. Immunol.* **2010**, *40*, 599–606. [CrossRef]
126. Liu, Y.; Li, S.; Zhang, G.; Nie, G.; Meng, Z.; Mao, D.; Chen, C.; Chen, X.; Zhou, B.; Zeng, G. Genetic variants in IL1A and IL1B contribute to the susceptibility to 2009 pandemic H1N1 influenza A virus. *BMC Immunol.* **2013**, *14*, 37. [CrossRef]
127. Zhang, G.; Zhou, B.; Li, S.; Yue, J.; Yang, H.; Wen, Y.; Zhan, S.; Wang, W.; Liao, M.; Zhang, M.; et al. Allele-specific induction of IL-1beta expression by C/EBPbeta and PU.1 contributes to increased tuberculosis susceptibility. *PLoS Pathog.* **2014**, *10*, e1004426. [CrossRef]
128. Lind, H.; Haugen, A.; Zienolddiny, S. Differential binding of proteins to the IL1B -31 T/C polymorphism in lung epithelial cells. *Cytokine* **2007**, *38*, 43–48. [CrossRef]
129. Luft, T.; Jefford, M.; Luetjens, P.; Hochrein, H.; Masterman, K.A.; Maliszewski, C.; Shortman, K.; Cebon, J.; Maraskovsky, E. IL-1 beta enhances CD40 ligand-mediated cytokine secretion by human dendritic cells (DC): A mechanism for T cell-independent DC activation. *J. Immunol.* **2002**, *168*, 713–722. [CrossRef]
130. Antonopoulou, A.; Baziaka, F.; Tsaganos, T.; Raftogiannis, M.; Koutoukas, P.; Spyridaki, A.; Mouktaroudi, M.; Kotsaki, A.; Savva, A.; Georgitsi, M.; et al. Role of tumor necrosis factor gene single nucleotide polymorphisms in the natural course of 2009 influenza A H1N1 virus infection. *Int. J. Infect. Dis.* **2012**, *16*, e204–e208. [CrossRef]
131. Garcia-Ramirez, R.A.; Ramirez-Venegas, A.; Quintana-Carrillo, R.; Camarena, A.E.; Falfan-Valencia, R.; Mejia-Arangure, J.M. TNF, IL6, and IL1B Polymorphisms Are Associated with Severe Influenza A (H1N1) Virus Infection in the Mexican Population. *PLoS ONE* **2015**, *10*, e0144832. [CrossRef] [PubMed]

132. Deval, H.; Alagarasu, K.; Mittal, M.; Srivastava, N.; Bachal, R.; Gondhalekar, A.; Chaudhary, U.; Chowdhary, D.; Bondre, V.P. Association of single nucleotide polymorphisms in TNFA and CCR5 genes with Japanese Encephalitis: A study from an endemic region of North India. *J. Neuroimmunol.* **2019**, *336*, 577043. [CrossRef] [PubMed]
133. Wang, B.; Wang, J.; Zheng, Y.; Zhou, S.; Zheng, J.; Wang, F.; Ma, X.; Zeng, Z.; Consortium, H.B.V.S. A study of TNF-alpha-238 and -308 polymorphisms with different outcomes of persistent hepatitis B virus infection in China. *Pathology* **2010**, *42*, 674–680. [CrossRef] [PubMed]
134. Glass, W.G.; McDermott, D.H.; Lim, J.K.; Lekhong, S.; Yu, S.F.; Frank, W.A.; Pape, J.; Cheshier, R.C.; Murphy, P.M. CCR5 deficiency increases risk of symptomatic West Nile virus infection. *J. Exp. Med.* **2006**, *203*, 35–40. [CrossRef] [PubMed]
135. Keynan, Y.; Juno, J.; Meyers, A.; Ball, T.B.; Kumar, A.; Rubinstein, E.; Fowke, K.R. Chemokine receptor 5 big up tri, open32 allele in patients with severe pandemic (H1N1) 2009. *Emerg. Infect. Dis.* **2010**, *16*, 1621–1622. [CrossRef] [PubMed]
136. Samson, M.; Libert, F.; Doranz, B.J.; Rucker, J.; Liesnard, C.; Farber, C.M.; Saragosti, S.; Lapoumeroulie, C.; Cognaux, J.; Forceille, C.; et al. Resistance to HIV-1 infection in caucasian individuals bearing mutant alleles of the CCR-5 chemokine receptor gene. *Nature* **1996**, *382*, 722–725. [CrossRef]
137. Zhou, J.; To, K.K.; Dong, H.; Cheng, Z.S.; Lau, C.C.; Poon, V.K.; Fan, Y.H.; Song, Y.Q.; Tse, H.; Chan, K.H.; et al. A functional variation in CD55 increases the severity of 2009 pandemic H1N1 influenza A virus infection. *J. Infect. Dis.* **2012**, *206*, 495–503. [CrossRef] [PubMed]
138. Chatzopoulou, F.; Gioula, G.; Kioumis, I.; Chatzidimitriou, D.; Exindari, M. Identification of complement-related host genetic risk factors associated with influenza A(H1N1)pdm09 outcome: Challenges ahead. *Med. Microbiol. Immunol.* **2018**. [CrossRef] [PubMed]
139. Ewulonu, U.K.; Ravi, L.; Medof, M.E. Characterization of the decay-accelerating factor gene promoter region. *Proc. Natl. Acad. Sci. USA* **1991**, *88*, 4675–4679. [CrossRef]
140. Zuniga, J.; Buendia-Roldan, I.; Zhao, Y.; Jimenez, L.; Torres, D.; Romo, J.; Ramirez, G.; Cruz, A.; Vargas-Alarcon, G.; Sheu, C.C.; et al. Genetic variants associated with severe pneumonia in A/H1N1 influenza infection. *Eur. Respir J* **2012**, *39*, 604–610. [CrossRef] [PubMed]
141. Ng, W.C.; Tate, M.D.; Brooks, A.G.; Reading, P.C. Soluble host defense lectins in innate immunity to influenza virus. *J Biomed Biotechnol* **2012**, *2012*, 732191. [CrossRef]
142. LeVine, A.M.; Hartshorn, K.; Elliott, J.; Whitsett, J.; Korfhagen, T. Absence of SP-A modulates innate and adaptive defense responses to pulmonary influenza infection. *Am. J. Physiol. Lung Cell. Mol. Physiol.* **2002**, *282*, L563–L572. [CrossRef]
143. Li, G.; Siddiqui, J.; Hendry, M.; Akiyama, J.; Edmondson, J.; Brown, C.; Allen, L.; Levitt, S.; Poulain, F.; Hawgood, S. Surfactant protein-A–deficient mice display an exaggerated early inflammatory response to a beta-resistant strain of influenza A virus. *Am. J. Respir. Cell Mol. Biol.* **2002**, *26*, 277–282. [CrossRef]
144. Hartshorn, K.L.; White, M.R.; Shepherd, V.; Reid, K.; Jensenius, J.C.; Crouch, E.C. Mechanisms of anti-influenza activity of surfactant proteins A and D: Comparison with serum collectins. *Am. J. Physiol.* **1997**, *273*, L1156–L1166. [CrossRef]
145. Herrera-Ramos, E.; Lopez-Rodriguez, M.; Ruiz-Hernandez, J.J.; Horcajada, J.P.; Borderias, L.; Lerma, E.; Blanquer, J.; Perez-Gonzalez, M.C.; Garcia-Laorden, M.I.; Florido, Y.; et al. Surfactant protein A genetic variants associate with severe respiratory insufficiency in pandemic influenza A virus infection. *Crit. Care* **2014**, *18*, R127. [CrossRef]
146. Yang, M.L.; Chen, Y.H.; Wang, S.W.; Huang, Y.J.; Leu, C.H.; Yeh, N.C.; Chu, C.Y.; Lin, C.C.; Shieh, G.S.; Chen, Y.L.; et al. Galectin-1 binds to influenza virus and ameliorates influenza virus pathogenesis. *J. Virol.* **2011**, *85*, 10010–10020. [CrossRef]
147. Chen, Y.; Zhou, J.; Cheng, Z.; Yang, S.; Chu, H.; Fan, Y.; Li, C.; Wong, B.H.; Zheng, S.; Zhu, Y.; et al. Functional variants regulating LGALS1 (Galectin 1) expression affect human susceptibility to influenza A(H7N9). *Sci. Rep.* **2015**, *5*, 8517. [CrossRef]
148. Hatesuer, B.; Bertram, S.; Mehnert, N.; Bahgat, M.M.; Nelson, P.S.; Pohlmann, S.; Schughart, K. Tmprss2 is essential for influenza H1N1 virus pathogenesis in mice. *PLoS Pathog.* **2013**, *9*, e1003774. [CrossRef]
149. Cheng, Z.; Zhou, J.; To, K.K.; Chu, H.; Li, C.; Wang, D.; Yang, D.; Zheng, S.; Hao, K.; Bosse, Y.; et al. Identification of TMPRSS2 as a Susceptibility Gene for Severe 2009 Pandemic A(H1N1) Influenza and A(H7N9) Influenza. *J. Infect. Dis.* **2015**, *212*, 1214–1221. [CrossRef]

150. Bartoszko, J.J.; McNamara, I.F.; Aras, O.A.Z.; Hylton, D.A.; Zhang, Y.B.; Malhotra, D.; Hyett, S.L.; Morassut, R.E.; Rudziak, P.; Loeb, M. Does consecutive influenza vaccination reduce protection against influenza: A systematic review and meta-analysis. *Vaccine* **2018**, *36*, 3434–3444. [CrossRef]
151. Belongia, E.A.; Skowronski, D.M.; McLean, H.Q.; Chambers, C.; Sundaram, M.E.; De Serres, G. Repeated annual influenza vaccination and vaccine effectiveness: Review of evidence. *Expert Rev. Vaccines* **2017**, *16*, 723–736. [CrossRef]
152. Leidner, A.J.; Murthy, N.; Chesson, H.W.; Biggerstaff, M.; Stoecker, C.; Harris, A.M.; Acosta, A.; Dooling, K.; Bridges, C.B. Cost-effectiveness of adult vaccinations: A systematic review. *Vaccine* **2019**, *37*, 226–234. [CrossRef]
153. Egli, A.; Santer, D.M.; O'Shea, D.; Barakat, K.; Syedbasha, M.; Vollmer, M.; Baluch, A.; Bhat, R.; Groenendyk, J.; Joyce, M.A.; et al. IL-28B is a key regulator of B- and T-cell vaccine responses against influenza. *PLoS Pathog.* **2014**, *10*, e1004556. [CrossRef]
154. Gelder, C.M.; Lambkin, R.; Hart, K.W.; Fleming, D.; Williams, O.M.; Bunce, M.; Welsh, K.I.; Marshall, S.E.; Oxford, J. Associations between human leukocyte antigens and nonresponsiveness to influenza vaccine. *J. Infect. Dis.* **2002**, *185*, 114–117. [CrossRef]
155. Linnik, J.E.; Egli, A. Impact of host genetic polymorphisms on vaccine induced antibody response. *Hum. Vaccin Immunother.* **2016**, *12*, 907–915. [CrossRef]
156. Poland, G.A.; Ovsyannikova, I.G.; Jacobson, R.M. Immunogenetics of seasonal influenza vaccine response. *Vaccine* **2008**, *26*, D35–D40. [CrossRef]
157. Burleson, G.R. Immunological Variation Due to Genetics of Inflammatory SNPs and Age and Impact on Disease Manifestation. *Toxicol. Pathol.* **2017**, *45*, 146–149. [CrossRef]
158. Castiblanco, J.; Anaya, J.M. Genetics and vaccines in the era of personalized medicine. *Curr. Genom.* **2015**, *16*, 47–59. [CrossRef]
159. Ellwanger, J.H.; Chies, J.A.B. Host genetic factors can impact vaccine immunogenicity and effectiveness. *Lancet Infect. Dis.* **2019**, *19*, 359–360. [CrossRef]
160. Zeng, M.; Nourishirazi, E.; Guinet, E.; Nouri-Shirazi, M. The genetic background influences the cellular and humoral immune responses to vaccines. *Clin. Exp. Immunol.* **2016**, *186*, 190–204. [CrossRef]
161. Allcock, R.J. The major histocompatibility complex: A paradigm for studies of the human genome. *Methods Mol. Biol.* **2012**, *882*, 1–7. [CrossRef] [PubMed]
162. Apanius, V.; Penn, D.; Slev, P.R.; Ruff, L.R.; Potts, W.K. The Nature of Selection on the Major Histocompatibility Complex. *Crit. Rev. Immunol.* **2017**, *37*, 75–120. [CrossRef] [PubMed]
163. Garred, P.; Genster, N.; Pilely, K.; Bayarri-Olmos, R.; Rosbjerg, A.; Ma, Y.J.; Skjoedt, M.O. A journey through the lectin pathway of complement-MBL and beyond. *Immunol. Rev.* **2016**, *274*, 74–97. [CrossRef] [PubMed]
164. Turner, M.W. The role of mannose-binding lectin in health and disease. *Mol. Immunol.* **2003**, *40*, 423–429. [CrossRef]
165. Tang, Y.W.; Li, H.; Wu, H.; Shyr, Y.; Edwards, K.M. Host single-nucleotide polymorphisms and altered responses to inactivated influenza vaccine. *J. Infect. Dis.* **2007**, *196*, 1021–1025. [CrossRef] [PubMed]

MDPI
St. Alban-Anlage 66
4052 Basel
Switzerland
Tel. +41 61 683 77 34
Fax +41 61 302 89 18
www.mdpi.com

Pathogens Editorial Office
E-mail: pathogens@mdpi.com
www.mdpi.com/journal/pathogens

www.ingramcontent.com/pod-product-compliance
Lightning Source LLC
Chambersburg PA
CBHW062100220926
51850CB00010B/2838

* 9 7 8 3 0 3 9 2 8 8 1 7 5 *